The Golden Age of the Great Passenger Airships

Graf Zeppelin & Hindenburg

Harold G. Dick with
Douglas H. Robinson

SMITHSONIAN BOOKS
WASHINGTON, D. C.

11 10 9 8 7   11 12 13 14 15
This book was edited by Ruth W. Spiegel and
designed by Alan Carter.
Type was set by BG Composition, Inc., Baltimore, Maryland.

Library of Congress Cataloging-in-Publication Data:

Dick, Harold G.
The golden age of the great passenger airships,
Graf Zeppelin and Hindenburg.

Includes bibliographical references and index.
1. Graf Zeppelin (Airship)—History. 2. Hindenburg
(Airship)—History. I. Robinson, Douglas Hill,
1918– . II. Title.
TL659.G7D53 1984 629.133′25 84-600298
ISBN 978-1-56098-219-7

Photographs not otherwise credited are from
Luftschiffbau Zeppelin, Friedrichshafen.

Cover: Photograph by Harold G. Dick of the *Hindenburg* being brought out for the second takeoff on the March 26, 1936, propaganda flight after temporary repairs had been made for damage incurred on the first takeoff. The *Graf Zeppelin* is seen overhead (Fig. 56).

The paper in this book meets the guidelines for permanence and durability of the Committee on Production Guidelines for Book Longevity of the Council on Library Resources.

Printed in the United States of America.

This book is dedicated to the memory of Dr. Hugo Eckener and Paul W. Litchfield, men of vision, who in the early 1900s had the foresight to see the possibility and capability of the great rigid airship in transoceanic commerce and transportation.

Contents

Preface

As the era of the big rigid airship receded into the past it became very evident that a close experience with the engineering design, construction, and operation of the big airships warranted recording the salient features of that fascinating era of flight. The period involved covers the years 1934 through 1938 when the *Graf Zeppelin* was making regularly scheduled passenger-carrying flights from Germany across the South Atlantic to Rio de Janeiro and the *Hindenburg* was making the first regularly scheduled passenger-carrying flights across the North Atlantic as well as nonstop flights from Germany across the South Atlantic to Rio.

The successful operation of the two great airships was followed by the disastrous burning of the *Hindenburg*. There followed the conversion of the *Hindenburg*'s sister ship to the use of helium instead of hydrogen with all the necessary arrangements for storing and purifying the helium. Then came the program for increasing the docking and operating facilities for a projected fleet of four great airships designed to carry passengers in luxury and comfort across the North Atlantic from Europe to the United States and return.

It was my good fortune as a young man to be involved in that fascinating period. I was uniquely qualified for my assignment to the Luftschiffbau Zeppelin by having an engineering degree from MIT, free balloon and blimp (nonrigid airship) licenses, and several years' experience in the actual design and engineering of the big rigid airships *Akron* and *Macon*.

Recording much of the information followed natural channels by way of my reports from Germany to Goodyear-Zeppelin and progress reports to President Paul W. Litchfield. In the later years, 1937 and 1938, as Germany became more involved with Nazism and rearmament, reports and technical information could not be forwarded to Goodyear but information could still be collected and retained in my personal notes. Because of the close and friendly relationship with the Luftschiffbau Zeppelin personnel, there was never any question concerning my accumulation of data.

This collection of data concerning the design, engineering, construction, and operation of the big rigid airships is the basis of this presentation. Material includes extensive notes, reports, and recorded data, all of which had to be edited and reduced to workable quantities, a monumental task.

The best qualified person to do this was Dr. Douglas H. Robinson, author of several books on the big rigid airships, who is recognized as the foremost historian concerning the big rigids not only in this country but in England and Germany as well.

My first contact with Doug Robinson was in the summer of 1937 when Doug, then nineteen years old, was bicycle touring through Europe and had stopped at Friedrichshafen to see the sister ship to the *Hindenburg*, the LZ 130, later named the *Graf Zeppelin II*. A call from Knut Eckener took me to Knut's office where I met Doug and then took him through the hangar showing him the LZ 130 which was then being modified to operate with helium inflation. I remember particularly Doug's repeated requests to be taken up into the ship but this was impossible—verboten—for safety reasons.

Years later our paths crossed again with Dr. Douglas H. Robinson taking on the chore of reducing the tremendous amount of information and data to workable proportions.

Much additional data and engineering information is available, too extensive and detailed to be included in this book. As it stands, I hope the story will interest even those who have never seen the giant rigid airships for it presents a participant's recollections of the years he devoted to building and flying them.

The opinions expressed in this book are my own.

Harold G. Dick

Acknowledgments

I am very grateful to Dr. Hugo Eckener who kept me informed of new and special developments and who made it possible for me to participate in so many flights of the *Hindenburg*, particularly the first test flight when I was the only person aboard not on the Luftschiffbau payroll.

The cooperation and assistance given by the engineering personnel of the Luftschiffbau Zeppelin and the operating personnel of the Reederei must be recognized. Without their willingness to share information this story would not be complete.

Dr. Karl Arnstein must be recognized for his help and support. His desire to have an engineering representative on board the *Graf Zeppelin* and *Hindenburg* as well as in close contact with the engineering and construction personnel at Luftschiffbau Zeppelin was of invaluable assistance during my stay in Friedrichshafen.

Special appreciation is due to Knut Eckener who not only offered information on the fabrication and erection of the *Hindenburg* and *Graf Zeppelin II* but also helped to make my stay in Germany more interesting by opening his home to me, inviting me to attend the Oktoberfest in Munich, and introducing me to some of the fascinating spots in the Friedrichshafen area.

Last, but not least, I am also grateful to Dr. Ross Taylor, professor and head of the English Department and head of the Department of American Studies at Wichita State University, who for many years insisted that I record in the greatest detail not only the engineering aspects of my five years in Germany, but also my association with Dr. Eckener, his son Knut, and other personnel of the Luftschiffbau Zeppelin. Without Dr. Taylor's insistence, this book would not have been written.

Note on the Photographs

In the period before World War II, both authors had access to the photographic archives of the Luftschiffbau Zeppelin, and procured many copies for their own use. It was assumed that the original photographic collection had been destroyed in the heavy bombing of Friedrichshafen during the war. In fact, the archives had been packed and dispersed in villages around the town, but were subsequently seized by the French, who occupied the area after the German surrender, and were taken away to an unknown destination. In 1955, Dr. Robinson learned that the City of Friedrichshafen was planning a small museum on the top floor of the city hall, and he wrote to the Bürgermeister, Dr. Grünbeck, inquiring about the photographs of the Zeppelin Company in view of his plans to use them in a published article. Dr. Grünbeck gave him permission to use any of the Luftschiffbau Zeppelin photos, but also requested that he send copies of the ones in his possession as the originals had disappeared. This was done. Subsequently a German official by accident discovered the archives of the Zeppelin Company at the Musée de l'Air in Paris, still in the original packing cases. After prolonged negotiation by the Bonn government, these photographs were finally returned to Friedrichshafen.

Introduction *The Nazis and Dr. Eckener*

he time is March 26, 1936. The place is Friedrichshafen in southern Germany. For slightly more than three years, Adolf Hitler has been the Führer and chancellor of Germany, and he and his Nazi followers have raised the German nation from the depths of despair and degradation to a level of pride and prosperity that they have not known for nearly twenty years. Gone is the shame and humiliation of being beaten in the Great War, and of being forced to admit their war guilt in the infamous Treaty of Versailles. Too proud and sensitive to admit that the German army of Frederick the Great, Blücher, and Moltke could be fairly and squarely beaten in the field, the German people have been looking for scapegoats—and Hitler has told them that they were betrayed by the Jews, the profiteers, the Social Democrats, and the November criminals who sought the Armistice at Compiègne. His plans for the future, as set forth in his book *Mein Kampf*, call for the elimination of all these enemies of the state, and since mid-1933 they have been disappearing into the concentration camps—Buchenwald, Dachau, Mauthausen—places whose real story will not be known until 1945. If necessary, the new government will rule by terror and intimidation: just how ruthless Hitler and the Nazis can be is revealed in the "Night of the Long Knives," June 30, 1934, when hundreds are murdered by the dreaded Gestapo. But the great mass of the German people are passive and indifferent to these events. There are jobs for all at last, many can and do parade in a uniform, and the build-up of the armed forces, in defiance of the Versailles Diktat, gives them back the feeling of pride and strength that they enjoyed under the kaisers.

But Hitler is not expanding the German armed forces simply to please the German people. He has long since planned to take back the territories forfeited to the enemy in 1919. Ultimately his rabid hatred of Communism points to a war against Russia—for lebensraum. The final goal is *Weltherrschaft*, the conquest of the world.

On March 1, 1935, a secret Luftwaffe, created in defiance of the Versailles Treaty, is revealed publicly by the Nazi air minister, Hermann Goering. On March 16, Hitler proclaims universal military service and a planned army strength of half a million men—five times the size of the 100,000-man Reichswehr permitted by the Versailles Treaty. Within a year there follows the first military aggression, when Hitler, on March 7, 1936, to the consternation of his generals, sends German troops across the Rhine bridges to occupy the demilitarized zones on the west bank. The Great General Staff fully expects that France, backed by England, will mobilize her vast army and chase their three battalions out of Aachen, Trier, and Saarbrücken.

May Day parade, 1935, of the "SA" (the Brown Shirts, *Sturmabteilung*) in Friedrichshafen. (Photo by Harold G. Dick)

But the Führer has correctly gauged the war-weariness of the French nation, and the readiness of English statesmen to accommodate the Nazis. Finding that the British will give them no support, the French government does nothing and thereby loses the best chance of stopping World War II before it begins. The German people are overjoyed, Hitler is triumphant, his generals discredited, and as bluff and intimidation bring more and more successes, not only Hitler himself, but even many army officers, see him as a military genius. To ratify his bloodless victory, Hitler proclaims a national referendum for March 29, in which the people can vote their approval of the occupation of the Rhineland. All the propaganda agencies of the Reich—Hitler is the first world leader to have in his cabinet a minister of propaganda, the shrill, satanic Joseph Goebbels—are turned on full blast to urge the German people to vote "ja" and support the Führer's policies. One of the propaganda minister's most spectacular and impressive instrumentalities is the Zeppelin airship, built by the Luftschiffbau Zeppelin (the Zeppelin Company) of Friedrichshafen am Bodensee and known throughout the world as a majestic symbol of the German Reich.

The love affair of the German people with the Zeppelin goes far back beyond World War I, and its creator, Ferdinand August Adolf, Count von Zeppelin, was an authentic folk hero long before Adolf Hitler went into politics. Where the imitators in Great Britain and the United States have found much skepticism and little support in their own countries, the German people over and over again have responded with almost religious fervor to the great ships hovering in the sky, *kolossal* testimony to the superiority of German science and technology, and even of the German spirit and culture. As compensation for the German national inferiority complex, the Zeppelin has no rival.

Although Zeppelin's first craft, flown in 1900, was 420 feet long and 38 feet in diameter—a giant for its day—LZ 1 and its successors were slow and carried a minimal load, and the Count was derided as the "crazy inventor." Paradoxically, it was an early disaster that made the Count an overnight national figure. Attempting to make a twenty-four-hour endurance flight in his fourth ship, in August 1908, Count Zeppelin found cheering crowds blackening the streets of German cities along his route down the Rhine to Mainz, and on the homeward course towards Friedrichshafen. But the twenty-four-hour requirement was not to be met. Engine failure caused a forced landing at Echterdingen near Stuttgart, and the LZ 4 was carried away by a squall and consumed by fire on striking the ground.

In an extraordinary outpouring of national feeling, the German people, demanding that the seventy-year-old Count continue his experiments, sent their contributions to Friedrichshafen, the total amounting to more than six million marks. Backed by the fervor of the masses, Count Zeppelin set up a passenger operation with airships carrying patriotic citizens for two-hour joy rides from hangars in the neighborhood of all the big German cities. Pressured by public opinion, the German Army took up the Zeppelin for long-range reconnaissance in the enemy's rear areas, and the German Navy tested the Zeppelin for scouting with the High Seas Fleet in the North Sea. Meanwhile the chauvinistic German press threatened potential enemies with the nightmare of aerial bombardment, and in England, phantom Zeppelin sightings at night were commoner than flying saucer reports in our day.

With the outbreak of World War I, the German people confidently expected that the Zeppelin—the first strategic bomber, and the unique possession of the German nation—would bring the conflict to a prompt conclusion by laying waste the enemies' cities and terrorizing their population to the point where their leaders would sue for peace. Eighty-eight Zeppelins were put into service by the German Army and Navy during the conflict, and although they flew bombing raids against the enemy until the last summer of the war, and suffered appalling casualties from defending aircraft which shot down many Zeppelins in flames, they did not bring the Allied governments to their knees. Fortunately for Count Zeppelin, he did not live to see the failure of his invention as a military weapon, for he died on March 8, 1917.

While the Count had seen the big airship as a weapon of air superiority in the next war, he had also realized that its enormous size, great endurance, and load-carrying capacity fitted it ideally for transoceanic passenger carrying. Proof of this contention came indirectly as early as the summer of 1919 when the British R 34, a copy of a wartime German Zeppelin, flew the Atlantic nonstop both ways with thirty people aboard while pioneer aviators, in grossly overloaded flying boats, were making their way painfully across the Atlantic for the first time—by short stages from New York to Newfoundland to the Azores to Portugal. But it was the Zeppelin Company which would prove the capability of the big rigid airship for ocean passenger service, and the credit would belong to one man, and one man only, Dr. Hugo Eckener.

Born in 1868 in Flensburg on the Baltic Sea, an enthusiastic deep water yachtsman and a student of social economics, Eckener became involved in the Zeppelin enterprise by accident. For reasons of health, he had moved to Friedrichshafen to write a dissertation on political economy, "Capital or Labor Shortage? A New Answer to an Old Question." Here he became interested in Count Zeppelin's early experiments, and in 1909 he joined the organization as a director of public relations. Presently Eckener found himself playing a more active role. After the first commercial Zeppelin came to grief on her first passenger flight, Eckener was invited to command the next ship, and shortly thereafter became flight director of the DELAG, the acronym for the Deutsche Luftschiffahrts-Aktien-Gesellschaft (German Aerial Transportation Company).

Finding himself in the unprecedented situation of having to establish policies, procedures, and operating regulations for the first passenger airline in the world, Eckener, an original thinker much experienced with wind and weather, insisted on thorough training for all flight personnel in theoretical aerostatics and meteorology, and constant practice in ship handling in all weather conditions; improved technology, particularly in respect to power plants; and establishment of an efficient aviation weather service. It was to Eckener's credit that between 1910 and 1914 the DELAG operated four Zeppelins on 1,588 flights, carrying 10,197 paying passengers, without a single death or serious injury.

With the outbreak of World War I and the abrupt expansion of the airship branches of the German Army and Navy, both services turned to the experienced personnel of the DELAG for their training needs. Eckener became the technical and operational adviser to Fregattenkapitän Peter Strasser, the German Navy's brilliant Leader of Airships, and was responsible for the training of more than fifty flight crews including over a thousand men. Thus, by the end of the war, Eckener had amassed an enormous amount of knowledge and experience in the technology and operation of the big airships. This, together with his massive intellect and forceful personality, made him the greatest airship commander of all time.

With a compelling vision of the Zeppelin airship as the future transoceanic passenger carrier, it was Eckener who succeeded the old Count in the control of the Zeppelin Company. But with the Versailles Treaty calling for the destruction of the Zeppelin works, and the London Protocols forbidding the construction of large airships, the Zeppelin Company's survival appeared problematical. It was Eckener who saved the day by securing a contract for a large airship for the United States Navy. With the future *Los Angeles* completed in the fall of 1924, the flight trials were deliberately planned by

Eckener to cover the length and breadth of the Reich. Once again the silver ship in the sky served as a focus for national pride and feeling. The transatlantic delivery flight to Lakehurst was a personal triumph for Eckener. His next concern was the construction of the famous *Graf Zeppelin*, designed to carry twenty passengers in great comfort. Her exploits under Eckener's command created the same enthusiasm in foreign countries as already existed in Germany.

The Nazi *Machtergreifung* (takeover) in January 1933 promptly involved the Zeppelin enterprise as it did every other institution of German life. On the one hand, it enabled Dr. Eckener to finance the construction of his ideal fifty-passenger transoceanic airship, the later *Hindenburg*. First Dr. Goebbels, the propaganda minister, contributed two million marks for the construction of the new ship. Then his rival, Air Minister Hermann Goering, offered nine million marks. But there was a price to be paid in return, as Eckener himself realized: the Zeppelin was forced to become a propaganda appendage of the Brown Shirt regime. Goebbels demanded that the Nazi flag—a black swastika in a white circle on a scarlet background—be presented in giant size from top to bottom of the airships' hulls amidships. The best compromise that Eckener could achieve was to have the swastika flags painted on the vertical fins. Then Goering forced Eckener to accept government control through the formation of the Deutsche Zeppelin Reederei, an operating company in which the national airline, Lufthansa, was the dominant partner.

Eckener, with his cosmopolitan, international outlook, his many contacts in America and other countries, and his deep loathing for the Nazi gangsters and their savage methods, resisted as best he could, but he dared not openly defy the new regime lest it imprison him and take over direct control of the entire Zeppelin organization. Thus, when Goebbels demanded that the *Graf Zeppelin* and the newly completed *Hindenburg* make a three-day flight in advance of the March 29, 1936, referendum, showering propaganda leaflets and making electioneering speeches via loudspeaker, Eckener was forced to agree—but it was inconceivable that he would lend himself to the Nazis' aims by commanding the big airship on this occasion. One senior captain of the Zeppelin Company, Ernst Lehmann, did not share Eckener's scruples, however, and was ready to do the Nazis' bidding as captain of the *Hindenburg*. Dr. Eckener could only reply that "if the airships were to be used for political purposes it would be the end of the airship."

Years before, when first appointed to command the second DELAG airship, Eckener, because an expectant crowd of spectators was present and he had a full load of influential passengers on board, attempted to bring the *Deutschland II* out of the hangar in a gusty crosswind. The ship got away from the ground crew and was wrecked, fortunately with nobody injured. Thereafter, Eckener consistently refused to take chances to satisfy the expectations of the public, and he required his subordinates to follow the same prudent and cautious procedures. But on the day of the propaganda flight, the Doctor was no longer in charge.

The problem of getting the *Hindenburg* out of the shed at Friedrichshafen-Löwenthal was rather unusual. The hangar was oriented in an east-west direction. Inside, the *Hindenburg*'s bow was to the west, but a gusty east wind was blowing at 18 mph. A downwind takeoff was necessary—a more critical maneuver than the usual takeoff into the wind.

Captain Lehmann had made many such downwind takeoffs. Though the flight was to commence at 4 a.m., he delayed for two hours hoping the wind would drop. Meanwhile, the *Graf Zeppelin*, which had had no difficulty getting away from the shed in nearby Friedrichshafen, had been cruising over the Löwenthal field. Despite the voiced opposition of Dr. Eckener, Lehmann resolved to risk the downwind takeoff rather than delay the propaganda flight. The *Hindenburg* was walked out of the shed, being held by the ground crew of several hundred men. Before all was ready, a preventer cable failed aft, and the stern started up while the bow was still held down by the ground crew around the control car forward. The forward car party could not hold the control car, which got away from them. The *Hindenburg* nosed up to 14 degrees, and the lower rudder and fin smashed against the ground. Temporarily out of control, the huge airship free-ballooned away

The lower fin of the *Hindenburg* was damaged at take-off on March 26, 1936, at what was to be the start of the three-day propaganda flight prior to the referendum on the annexation of the Rhineland. (Photo by Harold G. Dick)

from the field. Some emergency repairs were made and after a flight of 2 hours 53 minutes, the big Zeppelin made a normal landing.

Thanks to the sturdy design and construction of the *Hindenburg*, the damage was entirely local, confined to the lower after portion of the lower fin and the lower portion of the rudder. The bottom six feet of the damaged rudder was cut off and faired in, and the same was done with the lower fin. Later in the afternoon of March 26, the *Hindenburg* took off and departed from Friedrichshafen with the *Graf Zeppelin* on the three-day propaganda flight.

The consequences for Dr. Eckener and the Zeppelin concept were far more serious. Meeting Captain Lehmann on his return from the brief flight with the damaged tail, Dr. Eckener had angrily accused him in front of witnesses: "How could you, Herr Lehmann, order the ship to be brought out in such wind conditions? You had the best excuse in the world to postpone this idiotic flight; instead you risked the ship merely to avoid annoying Herr Goebbels."[1] When the propaganda minister heard of Eckener's remarks, he called a press conference and angrily announced: "Dr. Eckener has alienated himself from the nation. In the future, his name may no longer be mentioned in the newspapers, nor may his picture be further used."[2] Moreover, at the instigation of the Nazi government, Eckener was "kicked upstairs" to be chairman of the board of directors of the Zeppelin Company, while the control of operations of the Reederei, the transoceanic Zeppelin passenger line, was placed in the hands of Captain Lehmann, a man more acceptable to the Nazi leadership.

With Eckener's steadying hand lifted from the helm, the Zeppelin enterprise, like other agencies of the Nazi Reich, would now have to learn to live dangerously. The younger commanders—Max Pruss, Hans von Schiller, Albert Sammt—were not so conservative in their flying as was the prudent old master. At Lakehurst on the fatal May 6, 1937, Captain Pruss brought the *Hindenburg* up to the mast in a tight turn. At the subsequent inquiry, Dr. Eckener, a member of the German investigating commission appointed by Air Minister Goering, argued insistently that the sharp turn had overstressed the after hull structure, causing one of the bracing wires to break. The recoiling end, he theorized, slashed open one of the after gas cells, permitting the escape of a large amount of hydrogen which was ignited by the prevailing St. Elmo's fire observed atop the ship by at least two witnesses just before the hydrogen fire.

Thus, it is possible that the *Hindenburg*'s mishandled takeoff on the morning of March 26, 1936, the subsequent quarrel between Dr. Eckener and Captain Lehmann, and Eckener's banishment from the leadership of the Zeppelin Reederei may have led to the catastrophe at Lakehurst a year

[1]All notes are to be found at the back of the book, immediately following Appendix D.

A temporary repair was made to the *Hindenburg*'s lower fin. It was in this condition that the *Hindenburg* carried out the three-day propaganda flight. (Photo by Harold G. Dick)

later. Certainly the huge Zeppelin, with her crew of sixty-one and thirty-six passengers, would have been operated more prudently if that great airship-man, Dr. Hugo Eckener, had still been in command.

Among all the German participants and spectators of the first takeoff for the propaganda flight there was one American. He had brought his camera with him, and took photographs—the only ones that survive today—of the damaged fin. The Germans, particularly during the Nazi period, were excessively self-conscious about any public failure that undermined national prestige. Security guards and Zeppelin Company personnel raced around the field at Löwenthal, snatching away cameras and tearing out the film. But the American had concealed his camera inside his coat.

Who was this American? His name was Harold G. Dick, and he was present as liaison man between the Goodyear-Zeppelin Corporation of Akron, Ohio, and the parent firm, the Luftschiffbau Zeppelin. His adventures in Friedrichshafen between the years 1934 and 1938, his involvement with the building of the *Hindenburg*, his participation in flights to South America in the *Graf Zeppelin* and to South and North America in the *Hindenburg*, and his close association with Dr. Eckener and his son Knut are the subjects of this book.

Chapter 1 *How I Got Started with Airships*

I was seventeen years old when I entered the Massachusetts Institute of Technology in the fall of 1924, never thinking that so much of my life in the future would be intimately bound up with the big airships. Although MIT in 1913 had been the first institution of higher learning in the United States to offer a course in aeronautical engineering, I was aiming for a degree in mechanical engineering.

I was however attracted by the Army Air Service's ROTC program and signed up for this, though initially nothing much happened beyond expeditions to the old East Boston Airport, and rides around the area with National Guard pilots still flying World War I Jennies and de Havillands. Six weeks of summer camp at Langley Field in 1927 were something else: I got to occupy the nose cockpit of a twin-engined Martin bomber and, even more thrilling, had my first flight in an airship. This was an army blimp, a "rubber cow," one of the TC ships built by Goodyear (though I did not know it at the time). Underneath the fabric bag I rode with the crew in an open gondola suspended by a multitude of cables, with the two engines on outriggers attached to the car. It was fascinating, floating in the air—a different kind of flying from the small, fragile airplanes—and the memory stayed with me.

In 1929 I received my S.B. degree and a commission in the Air Service reserve. A year later, along with twenty-four other graduates of such prestigious institutions as Georgia Tech, NYU, and my own MIT, I was selected by Goodyear Tire and Rubber Company for a three-month staff training program; at the end we could choose the department to which we wished to be assigned. Knowing that Goodyear was constructing airships through its subsidiary, the Goodyear-Zeppelin Corporation, I felt some stirrings of the interest aroused by the flight at Langley Field, but decided in the end to go into mechanical goods design.

Here I enjoyed an early success. The ailing V-belt department was working only half a shift per day because of a lack of orders; the Goodyear product was inferior to the competition's. I helped to rewrite the company handbook on belting design and developed a new V-belt which was so successful that the department was shortly working three full shifts per day! Perhaps this incident brought me to the attention of Paul W. Litchfield, the president of the Goodyear Tire and Rubber Company.

While Goodyear, as a large manufacturer of rubber products, had been building pressure airships of rubberized fabric since 1911,[1] Litchfield had been impressed with the giant Zeppelin-type rigid airships, for which he saw a great future in both the military and commercial fields. Thus, in October 1923, he had entered into an agreement with Dr. Hugo Eckener, representing the Zeppelin Company of Friedrichshafen, whereby the Goodyear-Zeppelin Corporation of Akron, Ohio, would receive the North American rights to the Zeppelin patents, while the Luftschiffbau Zeppelin would receive not only German rights to any American patents, but also 10 percent of the stock in the American company. Furthermore, key Zeppelin Company personnel would come to America to convey the parent firm's expertise to Goodyear-Zeppelin. Thirteen engineers, headed by Dr. Karl Arnstein, chief stress analyst of the German firm, in fact arrived in America in October 1924.

In 1928, Goodyear-Zeppelin won an important U.S. Navy contract for the construction of two 6,500,000-foot rigid airships—later named the *Akron* and the *Macon*. By late 1930, not only was the design of the big ships going ahead, but the huge "Air Dock" had been completed and the construction of the *Akron* had gotten under way. At the same time, the Goodyear-Zeppelin design department was something of a German "closed shop," and Mr. Litchfield felt it was time that some of his bright young Americans should be added to the staff.

Dr. Karl Arnstein and his "twelve apostles," the original group of Germans who came to Goodyear as an engineering and design nucleus for the Goodyear-Zeppelin Corp. Here the group is shown aboard ship in October 1924. *From left to right, front row:* Schoettel, Schnitzer, Arnstein, Brunner, Klemperer; *center row:* Mosebach, Rieger, Liebert; *back row:* Bauch, Keck, Hilligardt, Helma, Fischer. (Photo courtesy Kane)

Thus it came about that one day late in 1930, three of us were summoned to Mr. Litchfield's office, where the great man told us of his desire to have some Americans involved in the rigid airship adventure. A charming and persuasive individual, Paul Litchfield argued that there was a great future with the development of the rigid airship, and it would be in our own interest to transfer to Goodyear-Zeppelin. For George Lewis and myself, "the oracle had spoken," and we promptly joined the Goodyear-Zeppelin staff.

Over the next four years, I worked in the Project Department, which was responsible for overall layouts of proposed airships, weights and performance calculations, and the like. More interestingly, I was admitted to a select group of about ten junior engineers with Goodyear-Zeppelin who were to be given overall experience and training to fit them in every possible way for responsible future positions in the design and construction of rigid airships. The lessons to be learned were not merely theoretical but also practical. After all, even the giant *Graf Zeppelin* was simply a powered balloon; hence we were taught practical aerostatics through free-balloon flying, then going on to train in the small Goodyear blimps, which at least gave us some idea of the problems of handling the huge rigids. I am still proud of the fact that my Fédération Aéronautique Internationale Spherical Balloon Pilot Certificate No. 1074, dated October 27, 1933, and my Dirigible Balloon Pilot Certificate No. 249, dated May 25, 1934, were signed by Orville Wright. The man who had made the first controlled powered flight at Kitty Hawk on that historic December 17, 1903, was still living in Dayton as the elder statesman of aviation!

Meanwhile, at the highest corporate level, to which I did not have access, the prescribed contractual relationship between Goodyear-Zeppelin and the parent firm, the Luftschiffbau Zeppelin, was not proceeding as expected. While Dr. Eckener, the chairman, was always a friend to President Litchfield and American airshipmen generally, at lower levels in the Zeppelin Com-

FÉDÉRATION AÉRONAUTIQUE INTERNATIONALE

NATIONAL AERONAUTIC ASSOCIATION OF U. S. A. INC.

Certificate No. 1074

The above named Association, recognized by the Fédération Aéronautique Internationale, as the governing authority for the United States of America, certifies that Harold G. Dick born 19th day of January, 1907 having fulfilled all the conditions required by the Fédération Aéronautique Internationale, for a Spherical Balloon Pilot is hereby brevetted as such.

Dated October 27, 1933

CONTEST COMMITTEE

Orville Wright, Chairman

Executive Vice-Chairman

[SEAL]

Signature of Licensee: Harold G. Dick

FÉDÉRATION AÉRONAUTIQUE INTERNATIONALE

NATIONAL AERONAUTIC ASSOCIATION OF U. S. A. INC.

Certificate No. 249

The above named Association, recognized by the Fédération Aéronautique Internationale, as the governing authority for the United States of America, certifies that Harold G. Dick born 19th day of January, 1907 having fulfilled all the conditions required by the Fédération Aéronautique Internationale, for a Dirigible Balloon Pilot is hereby brevetted as such.

Dated May 25, 1934

CONTEST COMMITTEE

Orville Wright, Chairman

Executive Vice Chairman

[SEAL]

Signature of Licensee: Harold G. Dick

Fédération Aéronautique Internationale (F.A.I.) dirigible balloon (blimp) and spherical balloon (free balloon) licenses. Most blimp flying was in the Goodyear *Defender.* Free ballooning was in an 80,000-cubic-foot balloon inflated with coal gas (mostly hydrogen). Both licenses are signed by Orville Wright. (Photo by Harold G. Dick)

Dr. Hugo Eckener, the dean of all airship men. He had no use for the Nazis and if he had not been such a world-famous figure he would most probably have been liquidated in the purge of the "Night of the Long Knives" in June 1934.

pany there was jealousy and resentment that Zeppelin "trade secrets" should be shared with the Americans. In particular, Dr. Eckener's ideal transoceanic airship, the 7,062,000-cubic-foot LZ 129, later christened the *Hindenburg*, was in the preliminary stage of construction, and Goodyear-Zeppelin was naturally desirous of obtaining full information about this craft, which represented an impressive advance over the *Graf Zeppelin* of 1928 vintage.

Dr. Eckener's response was to suggest that Goodyear-Zeppelin send to Friedrichshafen some of their blimp pilots to fly on the *Graf Zeppelin*'s scheduled passenger flights to South America. I knew nothing of this arrangement before early 1934 when some of us were meeting with Mr. Litchfield and one of Goodyear's vice-presidents, Joe Mayl, remarked that he understood that arrangements had been made with Dr. Eckener to send over some of our blimp pilots to fly in the *Graf Zeppelin*. Since fools will go where angels fear to tread, I spoke up and asked, "Why not send engineers who know something about flying and find out what it's all about?"

Mr. Litchfield changed the subject immediately but next day, I later learned, all hell broke loose. Dr. Arnstein, the chief designer at Goodyear-Zeppelin, immediately agreed: reports from a trained engineer on progress with the construction of the LZ 129 obviously would be worth more than the observations of the blimp pilots. Mr. Litchfield also felt the need to have a trustworthy and qualified representative of Goodyear-Zeppelin reporting to him from Friedrichshafen. Within exactly one week of my brash remark, George Lewis and I were told to pack our gear, as we were going to Friedrichshafen to fly with the Germans aboard the *Graf Zeppelin* and to follow the construction of the LZ 129. We were told by Dr. Arnstein to send him frequent technical reports on the progress of the new ship, while Mr. Litchfield wanted me to keep him informed on the general situation.

Thus, early in May 1934, George and I found ourselves aboard the SS *President Harding* en route to Europe. It would be fourteen months before I would see the United States again. I knew I would enjoy the direct support and concern of President Litchfield, as shown, for example, in a handwritten letter that I still treasure:

> Sept. 16, 1934
>
> Dear Hal,
>
> Yours of 9-4-34 at hand, and Engineer Pilot Dick and Hell-of-an-engineer Lewis are hereby ordered to equip themselves with red flannels, and dig in for a cold German winter. About April 1st you are ordered to Akron, Ohio, U.S.A., fully trained and capable of designing and constructing rigid airships as well as operating them. If you qualify on this assignment, maybe you can convince the Washington Bureaucrats that an airship is not a *haufenmist*[2]; if not, the wood shed for you both. Given under our hand and seal, this cool September morn, The Goodyear-Zeppelin Corp.
>
> by *P. W. Litchfield*
> *Pres. GZC*

Chapter 2 *The Dream of the Transatlantic Passenger Airship*

he Zeppelin Company of Friedrichshafen, which was my destination, was the original home of the rigid airship, and of the 161 of these craft ever built and flown, 119 were constructed here. Hence, Friedrichshafen was the mecca to which all true believers looked for inspiration and guidance, and the Luftschiffbau Zeppelin, led by Dr. Eckener, was the fount of all knowledge on the subject.

While most of the products of the Friedrichshafen firm had been built for the German Army and Navy in World War I, even before Count Zeppelin flew the LZ 1 in 1900 there had been the dream of connecting the continents by air. As early as 1910 the DELAG, a passenger-carrying subsidiary of the Zeppelin Company, was flying passengers in airships. These crude Zeppelins were incapable of transoceanic service. Amidships was an enclosed cabin for twenty to twenty-four passengers, with large windows which could be opened in warm weather, comfortable wicker chairs and tables, ornate mother-of-pearl inlays on the beams and pillars, and a washroom and pantry aft. The DELAG did not even attempt scheduled intercity service inside Germany, but found patriotic burghers eager to pay the equivalent of $50.00 apiece for two-hour joy rides in the vicinity of the airship sheds near the large cities. Several of the early passenger craft were lost after only a few weeks in service, but the later ships carried an impressive number of persons. Yet this was essentially a publicity operation designed to influence the German armed forces to purchase the craft for military purposes, and to maintain in the public the "Zeppelin fever" aroused by the 1908 tragedy at Echterdingen, where the Count's fourth airship had burned while trying to make a twenty-four-hour flight.

Under the stimulus of war, enormous advances in airship design and technology had been made. It was the German Navy which, to meet the needs of the High Seas Fleet for large and efficient scouting and raiding craft, had demanded that the Luftschiffbau Zeppelin hire outstanding engineers such as Dr. Karl Arnstein, while laying down the specifications for even larger well-streamlined craft, with greater and greater load-carrying capacities. In 1917 a modified North Sea type of Zeppelin, the L 59 of 2,418,700 cubic feet, set out from Bulgaria with 15 tons of cargo for German troops still fighting in East Africa. She was recalled over Khartoum and returned to her base after covering 4200 miles in ninety-five hours. On the drawing board at the end of World War I was the L 100, of 3,813,500 cubic feet, slightly larger than the later *Graf Zeppelin*, and fitted with ten Maybach "altitude motors" and with a calculated useful load of 92 tons.

With the signing of the Armistice ending World War I, the Luftschiffbau

Count Ferdinand von Zeppelin, the airship pioneer who gave his name to the Zeppelin type of rigid airship construction. Not all rigid airships were Zeppelins, those of the Schütte-Lanz Company being of quite different construction.

The LZ 1, Count Zeppelin's first airship, made its first flight on July 2, 1900, from a floating dock on Lake Constance (the Bodensee). The ship was trimmed by cranking the suspended weight between the gondolas fore and aft.

The LZ 10, *Schwaben,* a pre–World War I passenger-carrying airship. Between June 1911 and June 1912, this ship made 218 successful flights, carrying in all 4358 passengers. The *Schwaben* was one of four similar airships operated by the DELAG.

Zeppelin anticipated applying all the technological advances of the war years to building large airships for transoceanic service. Several designs were drawn up, and the company was quite serious in its intentions respecting the LZ 125, a reworking of the wartime L 100. Seven hundred seventy-four feet long, 98 feet in diameter, with 3,532,000 cubic feet of hydrogen in sixteen gas cells, and with twelve 240-HP Maybach engines, this "America ship" was offered in two versions: a mail and fast freight carrier with a few passenger cabins in the keel above the control car, and a passenger ship with a large cabin (with staterooms) forward and the control car in the nose of the gondola.

The Zeppelin Company wanted to develop public interest in airship travel, and while building an "America ship" with its own money would have been a reckless gamble both politically and financially, its directors resolved, only two and a half months after the Armistice, to construct a small passenger Zeppelin to make scheduled flights between Friedrichshafen and Berlin. This little craft, christened *Bodensee*, was of only 706,200 cubic feet volume, but on the short 370-mile flights she carried twenty to twenty-six passengers. A totally new design, the *Bodensee* reflected the best thought of the Luftschiffbau Zeppelin engineering staff on passenger carrying, and was far superior to the prewar DELAG ships. The passengers were accommodated forward in a streamlined gondola some 80 feet long built onto the hull, with the control car being in the forward portion. Inside the gondola were five compartments each seating four people. Wide windows offered an enthralling view of the scenery below, and a steward served light meals, fine wines, and liqueurs. A lengthened sister ship, the *Nordstern*, intended for a service to Stockholm, was completed but never flew under German colors. Instead, at the end of 1919, both were seized by the Inter-Allied Aeronautical Control Commission and awarded to Italy and France. Nonetheless, the successful design set the pattern for later and larger commercial airships.

The victorious Allies were determined to destroy the German airship industry as a competitor to their own ambitious plans for world airship commerce. The Versailles Treaty called for the destruction or surrender of all Zeppelins, the demolition of military hangars and gas works, and the razing of the building hangars and airship plant. This would have meant the end for the Luftschiffbau Zeppelin, but Dr. Eckener won a reprieve by interesting the U.S. Navy in procuring a craft of new and advanced design. In return, the United States government saw to it that the building hangars and facilities in Friedrichshafen were spared, at least until completion of the LZ 126. The Conference of Ambassadors of the Allied Powers, who authorized the U.S. Navy's agreement with the Zeppelin Company, limited the size of the ship to 2,500,000 cubic feet. Furthermore, it was stipulated that she be employed for "civil purposes," and hence necessarily was to be a passenger-carrying type.

Faced with the possibility that the LZ 126 might be the last Zeppelin to be built, Dr. Eckener and his staff did their best to create a sound and advanced

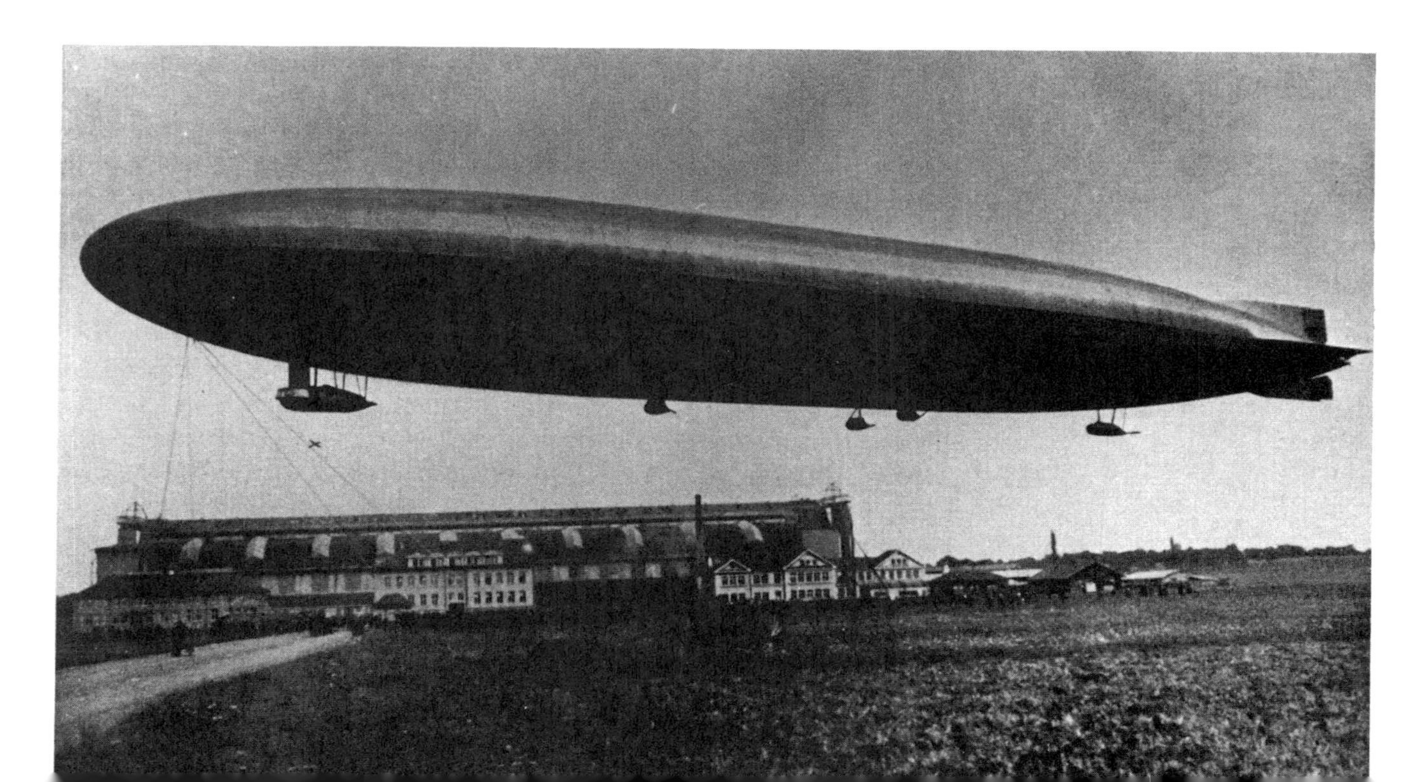

The L 71, last of the World War I military airships, shown at the airship works in Friedrichshafen. It first flew on July 29, 1918, and in 1920 was delivered to England.

The *Bodensee,* built in 1919 after World War I and before the Allies put a limitation on the building of airships by the Germans. This ship carried twenty to twenty-six passengers on its 370-mile flights between Friedrichshafen and Berlin.

The passenger quarters in the *Bodensee.* There were five compartments with wide windows. Light meals and refreshments were served in flight.

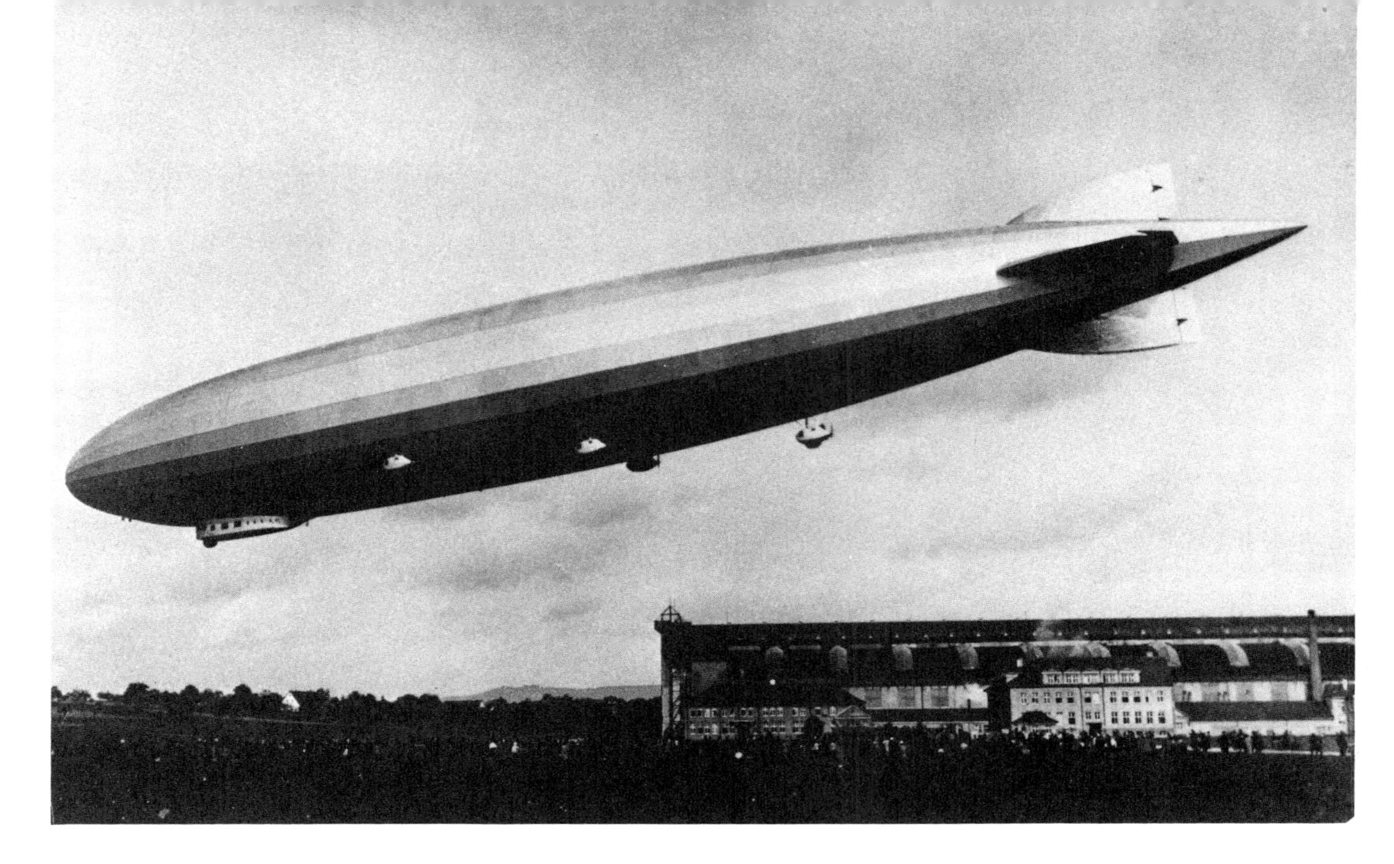

The LZ 126, built in Germany for the U.S. Navy, later named the *Los Angeles:* this was the first airship designed and constructed for trans-oceanic flights with passengers. It is seen here on a trial flight in Friedrichshafen in September 1924.

design for what would be their first transatlantic craft, even having a new 400-hp V-12 engine, the VL-1, created for her at the Maybach plant. Five of these engines were carried in single gondolas, in pairs to port and starboard, and one aft in the centerline. As in the *Bodensee*, a long gondola forward housed the control car in the nose, while there were five compartments in the rear, each to carry four passengers, the seats being convertible into beds at night. There was a small but well-appointed electric galley, a radio room, and washrooms and toilets right aft. Thus, the LZ 126 was well equipped to carry twenty passengers in comfort for the several days necessary for a transatlantic flight. Such a flight she was required to make for delivery to the U.S. Naval Air Station at Lakehurst, New Jersey, and with Dr. Eckener in command, did so without unusual problems in eighty-one hours between October 12 and 15, 1924. Christened *Los Angeles*, the German ship performed successfully for eight years, after which she was laid up for reasons of economy, and finally scrapped in 1939.

It was the successful transatlantic flight of the *Los Angeles*, as well as the *Bodensee*'s passenger flights, that aroused keen interest among Americans generally. This led, as related earlier, to the creation of the Goodyear-Zeppelin Corporation, with Dr. Karl Arnstein and twelve other Zeppelin engineers forming the nucleus of the American company's design department.

The German-built *Los Angeles*, however, was not the last airship constructed by the Zeppelin Company. After the Conference of Ambassadors on May 7, 1926, lifted the restrictions of the Versailles Treaty on civil aviation in Germany, nothing more was heard of destroying the Zeppelin works, and

Dr. Eckener was free to build—if he could raise the money—the airship that he knew he would need to persuade German and American financiers to back a transatlantic airship line. When we arrived in Friedrichshafen in 1934, the great pioneering days of this craft, the *Graf Zeppelin*, were a thing of the past: she had been making scheduled passenger flights to South America since the autumn of 1931. I was to see a great deal of this world-famous airship, and to enjoy the priceless experience of making six round-trip flights to Rio de Janeiro in her.[1] Thanks to Dr. Eckener's insistence that I be admitted everywhere to the activities of the Luftschiffbau Zeppelin, I was treated almost as a member of the crew, being assigned to one of the three watches on the ship to observe in the control car, where I often handled the rudder and elevator wheels, and otherwise familiarized myself with every part of the ship and with the duties of each crew member.

Chapter 3 *The* Graf Zeppelin

he most famous airship ever built—possibly the most famous aircraft—was the *Graf Zeppelin.* In every sense she was Dr. Eckener's creation. It was he who conceived her as a transoceanic demonstration craft that, boldly handled in spectacular pioneering flights, would stimulate interest in the commercial airship both in Europe and in America; it was he who led the publicity campaign over a period of two years to raise the money needed to begin construction; it was his vision, skill, and unrivaled experience that enabled the *Graf* safely to carry out all her long flights, while the airships of other nations met disaster. It was Dr. Eckener who deserved and received the credit for the *Graf*'s exploits, becoming in the process one of the world's most famous public figures in the fifteen years between the *Los Angeles*'s transatlantic delivery flight in 1924, and the outbreak of World War II in 1939.

Throughout her life the *Graf*'s performance was hampered by just one problem—lack of money. In the beginning, Eckener knew he could not look to the government of the Weimar Republic for aid, particularly as its aeronautical advisors, former combat fliers already hard at work building a secret air force, saw no military utility for the "gas bag." Remembering Echterdingen, Eckener hoped the German people might contribute financially as they had done after the loss of the LZ 4 in August 1908. Setting up the Zeppelin-Eckener Fund, the Doctor, loyally supported by his subordinates on the 1924 transatlantic crossing—Hans Flemming, Anton Wittemann, Hans von Schiller, and Max Pruss—spent the next two years in an exhausting round of public appearances, receptions, and speeches on the flight to America and transatlantic Zeppelins in general. Two and a half million marks were realized, but four million were needed for construction. Nonetheless, the funds in hand sufficed for Eckener, a master of the fait accompli, to start work on the LZ 127 at Friedrichshafen. Later the German government was persuaded to give over a million marks to complete the work.

An upper limit to the size of the new craft (as for the L 100 of 1918) was set by the measurements of the big Factory Shed II at Friedrichshafen, which had been completed in the summer of 1916. Its "clear inner dimensions" were 787 feet long, 138 feet wide, and 115 feet high. The hull of the LZ 127 measured 775 feet long and 100 feet in diameter, while the overall height of 110 feet including the gondola bumpers brought her to within two feet of the hangar arches when being walked in or out. With a length/diameter ratio of 7.8/1, and a passenger gondola far forward to decrease overall height, the *Graf*'s appearance was not as esthetically attractive as that of the perfectly streamlined *Los Angeles*, of which the *Graf* was a stretched version.

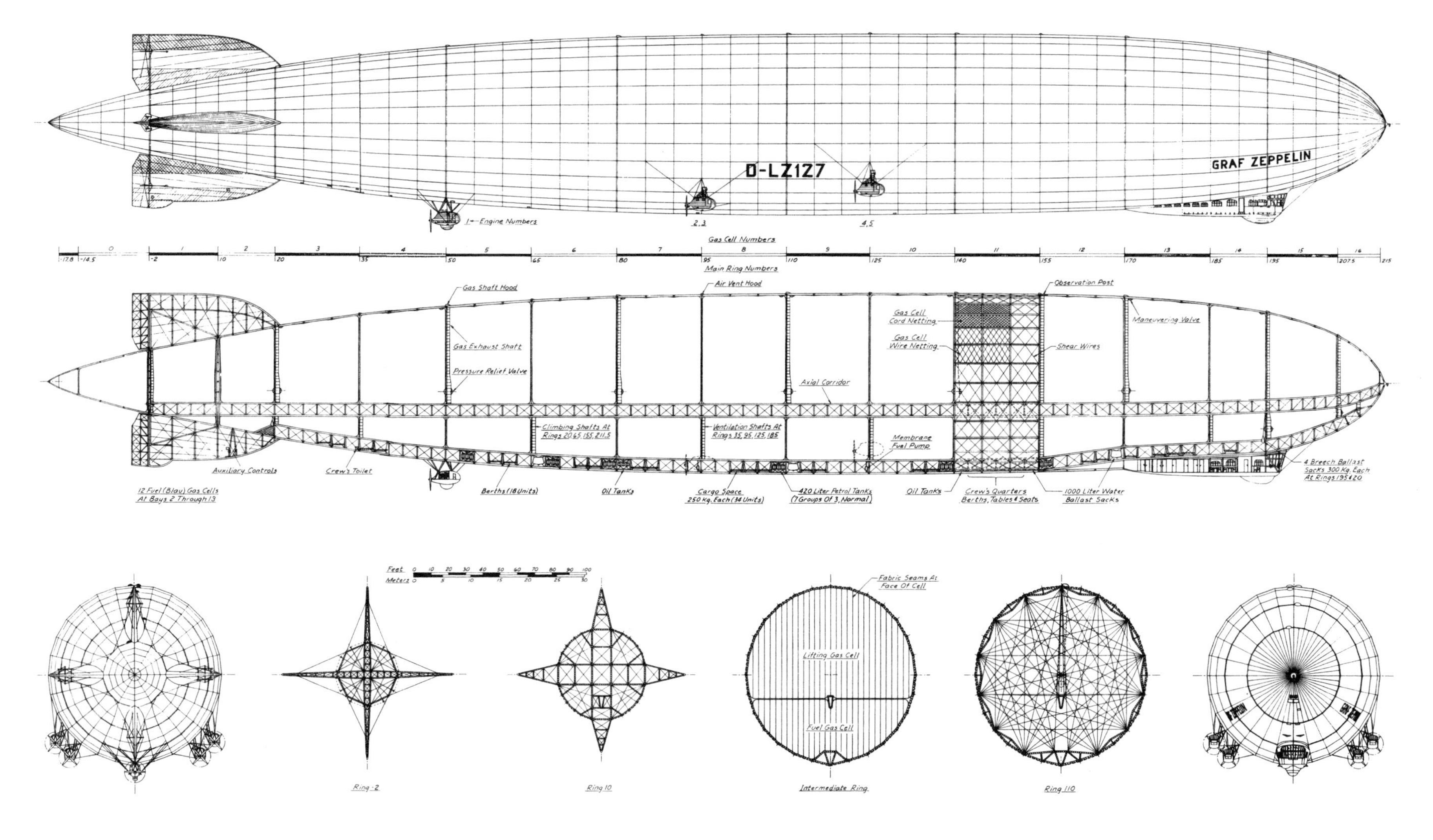

Side elevation, longitudinal section, and cross sections at designated rings or frames in the *Graf Zeppelin*. (Drawings by William F. Kerka)

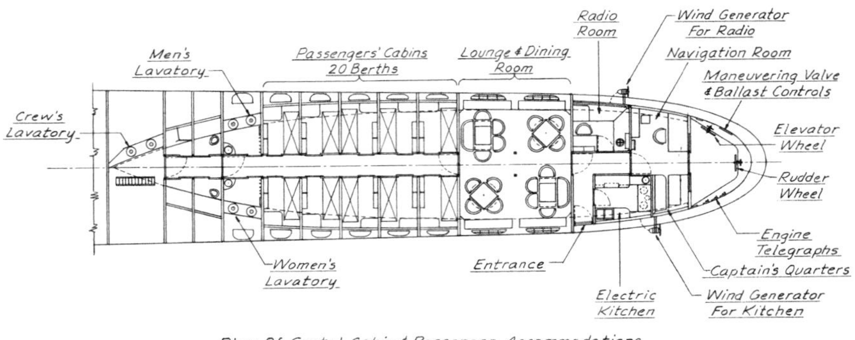

Plan of control cabin and passenger accommodations in the *Graf Zeppelin*. (Drawing by William F. Kerka)

A gas volume of 3,707,550 cubic feet made her the largest airship in the world at that time.

Main frames or rings were 15 meters apart, with two light intermediate frames spaced at a distance of 5 meters. The hull was divided by the main frames into seventeen compartments for gas cells. The fully inflated craft did not contain 3,707,550 cubic feet of hydrogen, however, for in twelve of the largest bays were *two* gas cells, the lower ones being designated as "fuel gas cells" with a maximum volume of 1,059,300 cubic feet. (In practice, about 750,000 cubic feet of fuel gas was loaded for the flights between Friedrichshafen and Pernambuco, and the remaining hull space filled to nearly 100 percent fullness with hydrogen.) The fuel gas, weighing approximately the same as air, was called "Blau gas" not because of its color (*blau* means "blue" in German), but for its developer, Dr. Hermann Blau, and was manufactured in the United States by the Union Carbide Company. By the time I reached Friedrichshafen, the shortage of foreign exchange was such that the Germans were making their own fuel gas by mixing hydrogen and propane. The rather impure propane has an unpleasant smell, like cooking cabbage, which permeated our clothing and attracted attention at indoor gatherings ashore!

It should be added that since the automatic valves of the upper hydrogen cells could not be reached from the keel in the bottom of the ship, a central gangway was installed—rather below the centerline of the ship—to enable the sailmakers to have access to these valves. The great advantage of carrying the Blau gas, with a specific gravity of 1.06, was that its consumption by the engines did not lighten the ship, and hydrogen did not have to be valved off wastefully during the flight to keep the ship in static equilibrium, as would have been the case if gasoline had been burned. Some gasoline was carried in tanks along the keel, however, and the actual combination of fuels used in practice is shown on the loading sheets for one of the South American flights in which I took part (*see* the data on Flight 366 in Chapter 8, *below*).

Aft were thick cantilever fins, with a cramped auxiliary steering station in the lower one. Forward was a mooring cone for attachment to mooring

The *Graf Zeppelin,* LZ 127, over the airship works in Friedrichshafen. The smaller hangar on the left is the docking shed of the *Graf Zeppelin,* while the larger shed houses the *Hindenburg,* then under construction.

masts. Dr. Eckener still preferred ground handling parties and a hangar, and *Graf Zeppelin* never attempted to moor at the high mast at Lakehurst. Beginning in 1931 with scheduled passenger flights to Rio de Janeiro, the *Graf* did regularly moor to a stub mast at the intermediate stop at Recife de Pernambuco. There were five engine gondolas, arranged as in the *Los Angeles*, each containing a Maybach VL-2 V-12 engine. This was a modified VL-1 with heavier crankshaft and higher compression ratio; it delivered 550 hp at full throttle, or 450 cruising at 1400 rpm.

The forward gondola, arranged much as in the *Los Angeles,* accommodated twenty passengers, but was larger—98½ feet long and 20 feet wide. Forward was the control room with rudder and elevator wheels, gas and water ballast controls, and engine telegraphs. Instruments included both a magnetic and a gyro compass for the rudder man, and altimeter, variometer,

The forward portion of the *Graf Zeppelin*'s control car. The rudder control is at the center, and the elevator control is on the left (port) side. Valve and ballast controls are also in this area. Engine telegraphs are at the upper right (starboard) side.

The aft power car of the *Graf Zeppelin*. This was mounted on the centerline of the ship forward of the lower fin. The engine was a Maybach VL-2 capable of developing 550 HP. Cruising power was 450 HP.

statoscope, inclinometers, and air and gas thermometers for the elevator man. In addition, his station had a new feature: indicators showing when the seventeen gas cells were expanded to 100 percent fullness.

Immediately abaft the control room was a chart room running the width of the gondola. Next to the rear was the radio room on the port side, and the galley to starboard. Both obtained their power from small wind-driven generators, which could be extended on brackets from the sides of the gondola. There was a 140-watt main transmitter and a 70-watt battery-powered emergency transmitter and receiver, together with a radio direction finder. The galley had a stove with two electric burners, a hot water heater, refrigerator, racks for silver and dinnerware, and the like. The door for entering the long gondola was immediately abaft the galley on the starboard side.

Next came a lounge running the width of the gondola and measuring 16½ by 16½ feet, lighted by four large outward-slanting windows and furnished with four dining tables and sixteen chairs. To the rear, five pairs of sleeping cabins opened off a corridor, providing accommodations for the twenty passengers. Each cabin had its own window and was furnished with a sofa, a clothes closet, a small table, and a folding canvas stool. At night, the back of the sofa would be hinged upward and attached to the overhead to form an upper berth. Right aft in the gondola were washrooms and toilets. Waste water, instead of being vented overboard, was retained and used as ballast.

On July 8, 1928, which would have been Count Zeppelin's ninetieth birthday, his only child, the Countess Hella von Brandenstein-Zeppelin, christened the ship with her father's name. Considerable detail work remained to be completed, however, and not until September did the *Graf* start her trials. After five test flights, the airship departed from Friedrichshafen for America on October 11, 1928, carrying forty crew members and twenty passengers. Six of these passengers were reporters, among them Karl von Wiegand and Lady Grace Drummond Hay representing the Hearst press. Thus commenced a fruitful relationship with the powerful American newspaper king. William Randolph Hearst made the major financial contributions to many of the later flights, while the two reporters traveled many times in the *Graf* and later in the *Hindenburg*. The invited presence of so many newspaper people was a shrewd gesture on the part of Dr. Eckener, who well realized the value for the airship cause of generating favorable publicity not only among German and American people, but also with American capitalists who might be persuaded to finance a transatlantic operation.

At the same time, newspaper readers on both sides of the Atlantic were hungry for articles on the progress of aviation and the daring exploits of the pioneer airmen. Today literally hundreds of persons cross the Atlantic daily in giant jet aircraft—flying near the speed of sound—with such regularity that this is no longer news. In 1928, less than ten years had passed since the

The lounge and dining salon of the *Graf Zeppelin.*

first crossing of the North Atlantic by air, and Lindbergh's solo flight from New York to Paris, which created intense enthusiasm for flying throughout America, had taken place only sixteen months before. Now at last twenty passengers were being carried by air across the ocean in Pullman-car luxury and comfort, expecting to arrive in America in less than four days—surely an epoch-making event that would be front-page news until the *Graf* arrived safely at Lakehurst.

All went well as the Zeppelin proceeded through the Straits of Gibraltar and on to Madeira. Early in the morning of October 13, a vicious-looking blue-black squall line was sighted advancing rapidly from the northwest. An inexperienced elevator man allowed the airship to pitch up violently at an angle of 15 degrees in passing through the squall line, and while there was no damage to the framework, the bottom cover of the port stabilizer was torn off by the extreme aerodynamic forces. Dr. Eckener sent out a message requesting assistance, but repairs were made by crew members and the ship went on. Another squall line west of Bermuda on the evening of October 14 caused a small tear in the upper covering of the port fin. The *Graf*'s narrow escape from disaster made headlines in America, and Eckener now capitalized on his fame by showing off the ship in a triumphal procession over Washington, Baltimore, Philadelphia, and New York before landing at Lakehurst. Repairs to the damaged fin took twelve days, and it was not until October 31 that the big airship was back in Friedrichshafen. The double

Atlantic crossing with passengers was a sensation, but to Dr. Eckener, it proved what he already knew—that the *Graf* was too small and too slow for regular North Atlantic service. In her eight-year existence the *Graf Zeppelin* crossed the North Atlantic only seven times.

An even greater sensation was the *Graf*'s world flight of 1929—only the second circumnavigation of the globe by air ever. (The first was the 1924 journey of two U.S. Army Air Service planes which took 175 days to fly the route with sixty-nine stops.) Since William Randolph Hearst had paid $100,000 to finance the *Graf*'s flight on condition that it commence at the Statue of Liberty in New York Harbor, the *Graf* came west to Lakehurst on August 5, 1929. Two days later she headed east with twenty-two passengers including the U.S. Navy's Commander Charles E. Rosendahl, commander of the *Los Angeles,* Lieutenant Jack Richardson, Lady Drummond Hay, and the Arctic explorer Sir Hubert Wilkins. After a five-day layover in Friedrichshafen, the big ship went on with twenty passengers, including German reporters and representatives of the Russian and Japanese governments. The route took them far north to follow the Great Circle course across Siberia, causing the passengers to marvel at the contrast between the dismal, empty swampland below and the comfort and luxury of the *Graf Zeppelin*'s cabins.

Rising to 6000 feet to clear the Stanovoi Mountains, the *Graf* went on to Tokyo, where she was housed in a former German airship shed re-erected at the Japanese naval air station at Kasumigaura. The passage from Friedrichshafen had taken 101 hours 49 minutes. On August 23, the *Graf* set off for the west coast of the United States, making a dramatic entry over San Francisco's Golden Gate at the magic hour of 4 P.M. on August 25, and then flew down the California coast to land in Los Angeles the following morning. There were some tense moments on the evening of August 26 when the ship failed to rise on lift-off because of an inversion, and had to hurdle a power line through the use of the elevators.

The Kurgarten Hotel, Dick's home away from home, in Friedrichshafen. This building was at one time the home of Count Zeppelin, who maintained a suite of rooms there. It was burned out and gutted by bombs in World War II and had not been rebuilt when Dick was in Friedrichshafen in 1967. (Photo by Harold G. Dick)

The flight to Lakehurst thereafter was uneventful, and on September 4 the *Graf* returned to Friedrichshafen. The "American" world flight had taken 12 days 11 minutes flying time. The sensation and enthusiasm engendered by Dr. Eckener's dramatic circling of the earth, with twenty-odd passengers for whom the pioneering journey was also a pleasure cruise, aroused the interest of American financiers. All of this might well have led to a transoceanic airship line with American participation—if the stock market had not crashed less than two months later.

Between May 18 and June 6, 1930, the *Graf* made her first flight to South America, stopping first at Seville to pick up a party of Spaniards including the Infante Don Alfonso de Orleans, a cousin of the king of Spain. Then the airship proceeded via the Canary Islands, the Cape Verdes, Saint Paul's Rocks, and Fernando de Noronha, to Recife de Pernambuco on the northeast bulge of Brazil. For two days the big ship lay at the stub mast and then flew on to Rio de Janeiro. Here the stay was brief, as there were no mooring facilities, and she returned to Recife, where she lay for two days more. Instead of returning to Friedrichshafen, however, the *Graf* went on to Lakehurst, where she was led into the hangar on the traveling stub mast developed by the Americans for the *Los Angeles.* En route home via the Azores, the Zeppelin landed again at Seville to off-load the Spanish guests and then went up the Rhone valley to Friedrichshafen.

The last of the spectacular pioneering flights was made in 1931, when the *Graf,* with a party of twelve scientists from four different nations, plus two reporters (one of them the novelist Arthur Koestler), journeyed via Leningrad into the Arctic. During the five days from July 26 to 30, the airship proceeded to Franz Joseph Land and a water landing to exchange mail with the Russian icebreaker *Malygin* for the benefit of stamp collectors, who in these years provided much of the financial support for the flights of the *Graf Zeppelin.* Then, while meteorological observations were made with radiosondes carried aloft by balloons, and the Earth's magnetic field was explored with sensitive instruments, the *Graf* flew down the west coast of remote Severnaya Zemlya, mapping this unknown territory with a panoramic camera. There followed exploration of the Siberian coast from Cape Chelyuskin to Dickson Haven, and further mapping of Novaya Zemlya, before the return to Leningrad. Apart from the scientific results, the flight was another spectacular publicity triumph.

After the cabin furnishings were reinstalled, the *Graf* made three scheduled passenger flights to South America in the fall of 1931. Because there were still no terminal facilities in Rio, the flights ended at Pernambuco and passengers were flown to the Brazilian capital via the German Condor airline. The success of these voyages persuaded the Zeppelin Company to concentrate on the South American service. In 1932 there were nine passenger crossings, and in 1933 there were nine more. President Vargas of Brazil was a guest on a flight from Pernambuco to Rio, and agreed to assist with the building of an airship base with hangar at Santa Cruz, some twenty-five miles south of the capital.[1]

The Nazi assumption of power in January 1933 had prompt repercussions for the *Graf* and for the Zeppelin Company. On May 1, the *Graf* appeared at the massed celebration by a million Brown Shirts on National Workers' Day in Berlin. By October the big ship, in accordance with German civil aviation regulations that applied to all aircraft, bore on the port side of her tail fins the Nazi colors—a black swastika in a white circle on a scarlet background—while on the starboard side there appeared the German national colors in horizontal stripes of black, white, and red. On a triangle flight that month, from Friedrichshafen to Rio to the Goodyear Air Dock at Akron, Ohio, Dr. Eckener was dismayed to find a new hostility toward the "Nazi ship." So strong was the feeling that when the *Graf,* by invitation, appeared at the Chicago World's Fair, Dr. Eckener ordered the rudder man to steer the airship in a clockwise circle around the city, so that those on the ground could not see the hated swastikas on the port side.

With the South America passenger flights a success, both operationally and financially, the Zeppelin Company planned twelve round trips for 1934 at two-week intervals, the first leaving Friedrichshafen on May 26, and the last on December 8. It was my great good fortune to participate in the second flight, departing from Friedrichshafen on June 9, 1934, and in five additional flights that summer and fall.

Chapter 4 *My Life in Friedrichshafen*

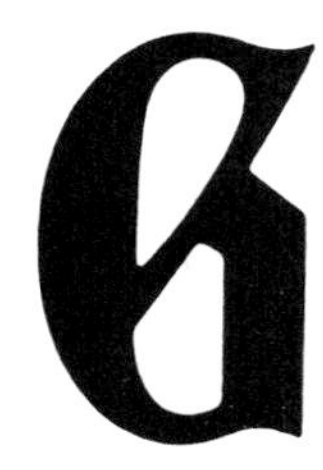

eorge Lewis and I landed at Le Havre on Friday, May 25, 1934. We traveled by train to Basel where we spent that night, and arrived in Friedrichshafen about noon on Saturday, May 26. Immediately George and Navy Lieutenant Commander George H. ("Shortie") Mills, who had been traveling with us, were notified that they would go to South America that night in the *Graf Zeppelin*. Since my turn would come with the second flight due to leave two weeks later, on June 9, I had time to settle in and get acquainted with Luftschiffbau Zeppelin personnel who would, I hoped, provide me with information on the new LZ 129 (later christened the *Hindenburg*) which at that time was in the early stages of construction.

Friedrichshafen, my home away from home for most of the next five years, had many attractions. Located on the shores of Lake Constance (the Bodensee in German), Friedrichshafen had facilities for a population of 10,000, but was in 1934 inhabited by some 28,000 people. It was situated on the north side of the lake, with the Swiss mountains across the lake to the south, the snow-capped Säntis peak rising 8215 feet high, 2200 feet of it above the timberline. Later I was to become intimately acquainted with the mountains, and would spend many weekends climbing in them. The upper Rhine, forming the western boundary of the tiny 61-square-mile principality of Liechtenstein, enters Lake Constance on the Swiss side at the southeast corner, and leaves the 38-mile-long lake at its southwest end on its way to Basel.

The towns around the lake are old and picturesque. On the German side there is Constance (Konstanz) with its famous Insel Hotel, the wood-carving town of Überlingen, and Meersburg with its old castle dating back to the seventh century. Manzell, just west of Friedrichshafen, is where Count Zeppelin had built his first six airships between 1900 and 1909. It was in 1909 when, at the invitation of the city fathers, he had moved his enterprise to the location where I found it on the north side of Friedrichshafen. Not only were there two huge hangars, the one built in 1916 where the *Graf* had been constructed, and an even larger one completed in 1930 where the LZ 129 was now coming along, but also the Maybach engine plant and the Zahnradfabrik, or gear manufacturing works. The Manzell facility was occupied by the Dornier airplane works.

Then there are the picturebook towns of Wasserburg and Lindau, the island resort at the east end of the lake. The only town on the Austrian side is Bregenz, famous for the Pfänderbahn to the top of the neighboring mountain with a great view of the lake. On the Swiss side is Altenrhein where the Rhine river enters the lake, Romanshorn directly across the lake

The Zeppelin fountain. Both the Rathaus and the fountain were destroyed by bombs in World War II. (Photo by Harold G. Dick)

from Friedrichshafen (with a connection by small ferry steamers), and Kreutzlingen, the Swiss part of the town of Constance. Between the towns are fruit-growing areas with some vineyards, mostly on the German side with southern exposure on the hillsides.

Our first stop in Friedrichshafen was the Kurgarten Hotel, owned and operated by the Luftschiffbau Zeppelin and at that time the finest hotel in southern Germany. As soon as possible George and I established quarters with a German family in some rooms that rented for 80 marks per month for the two of us, 10 marks for service, 0.75 marks each for breakfast, but we had to buy our own coal.

One problem was that I was dealing every day with the Germans, and while the educated members of the Luftschiffbau Zeppelin staff spoke fair to good English, it was desirable that I should learn to talk their language. I had had two years of German in high school in Lawrence, Massachusetts (my teacher's name was O'Leary!), but while George Lewis and I were together during my first tour in Friedrichshafen from May 1934 to July 1935, we naturally spoke English. When I returned alone for the second tour, from October 1935 through June 1936, it was obviously necessary for me to become proficient in German as I was now alone and there was nobody to whom I could speak English. Accordingly I contracted to take three German lessons a week from a twenty-three-year-old fräulein who taught at the Paulinenstift, a local convent. Languages have never been my cup of tea and I had to learn to laugh at myself, for I made some ridiculous mistakes in the beginning. My accent also baffled many Germans. Friedrichshafen being situated on Lake Constance, where the *Länder* of Baden, Württemberg, and Bavaria come together, I picked up some of the Schwäbisch accent and dialect. This, combined with my New England accent which in some ways is similar to the Low German (Plattdeutsch) of north Germany adjacent to Holland, made it hard for my listeners to tell whether I came from the north or south!

The Rathaus and the Zeppelin fountain in Friedrichshafen. (Photo by Harold G. Dick)

While George was participating in the *Graf Zeppelin*'s first transatlantic flight of the 1934 year, I attempted to get acquainted with the Zeppelin Company personnel to whom I would have to look for the information that Mr. Litchfield and Dr. Arnstein had sent us to Germany to get. My contacts for engineering and design covered the entire field, starting with Dr. Ludwig Dürr, who had joined Count Zeppelin's staff in 1899 and had been chief designer since 1901. In 1934 he was fifty-six years old and headed up all design and construction under Dr. Eckener, who was the chairman of the company. My other contacts were Albert Ehrle in structural design, Arthur Foerster for stress analysis, Fritz Sturm in power plants, Erich Hilligardt for instruments and electricity, Max Schirmer for wind tunnel and propellers, and Alfred Kolb for ground handling equipment. Gas cells were produced at the Ballon-Hüllen-Gesellschaft at Berlin-Tempelhof, and my contact there was Herr Strobl. By 1938 Foerster, in addition to heading up stress analysis, also had the responsibility for structural design. All of these men were extremely cooperative and did not hesitate to answer questions and release information. The association with them was pleasant.

I soon felt very much at home in Friedrichshafen. Being a bachelor, I did not feel too homesick for the United States, and for diversion I had the nearby towering mountains of Switzerland and Austria to climb or ski in each weekend in the company of German friends.

Chapter 5 *Knut Eckener and His Father*

uring the summer of 1934 I had the opportunity to become well acquainted with Knut Eckener, Dr. Eckener's son, who was approximately my age.[1] He was a *Diplomingenieur*; as far as I could determine this would correspond to having a master's degree in engineering in our country. Knut was our prime source of information on the construction and assembly of the LZ 129. We soon found that we liked many of the same things and had much in common even beyond our vital interest in airships. Between flights of the *Graf,* Knut would take me to many of the interesting spots around Lake Constance and to the choice places where the Germans would go for their "coffee and cake" on a Sunday afternoon or for the best beer in the area along with the famous raw smoked ham. Our lunch at times was nothing more than good Bavarian beer, rye bread, and sausage, which we both enjoyed.

During my second stay in Friedrichshafen my contacts with Knut became even closer. Knut's responsibilities during the period from 1935 to 1938 were increased considerably. By 1938 he had direct responsibility for all fabrication, assembly, and erection of the airships as well as all production of subcontract work that was being done elsewhere. The only production that did not fall under Knut's supervision was in the machine shop.

In 1938 I received the finest compliment from Knut, and I am sure he would not have paid me this honor without first having discussed the subject with his father and the leadership of the Luftschiffbau Zeppelin. Knut had charge of airship construction and erection, and his assistant, who shared Knut's office, was about to leave. Knut suggested that I consider leaving Goodyear and take the position as his assistant, sharing his office with him and the supervision of their airship construction. Although I was highly complimented, this position was something I could not accept.

The dominant figure in Friedrichshafen was Knut's father. During the almost five years I spent in Friedrichshafen, particularly in the latter years, I became very well acquainted with Dr. Eckener. He was the head of the airship construction company, the Luftschiffbau Zeppelin, and when an operating company was organized in 1935, the Zeppelin Reederei, he was the head of both companies. Captain Lehmann of the operating company and Dr. Dürr of the construction company both reported directly to Dr. Eckener. After Captain Lehmann's death in the *Hindenburg* fire, Herr Issel took over as managing director of the operating company.

Although Dr. Eckener made some of the flights with us in 1934 to Rio aboard the *Graf Zeppelin*, it was not until 1935 and 1936 when the *Hindenburg* started flying that I really got to know the Doctor, or "the old man," as

Knut Eckener on the scaffolding at the stern of the *Hindenburg*. The stern originally was to have been more pointed and seven feet longer, but it was shortened and rounded so that the *Hindenburg* could be docked in the Lakehurst hangar.

Commander Charles Rosendahl and Harold Dick at the dedication of the new Luftschiffbau Zeppelin Museum in Friedrichshafen on July 8, 1938. (From the collection of Harold G. Dick)

he was affectionately referred to. He had apparently found that I enjoyed the German people and their customs and that I liked to get out in the mountains. My behavior with respect to the political situation in Germany must also have met with his approval. He was not in the good graces of the Nazis and had little respect for Hitlerism.

His falling out with the Nazis had taken place some time before I was sent to Germany. In the summer of 1932, he had been approached by the future *Kreisleiter* (regional leader) of the Friedrichshafen area with a request for the use of the big construction hangar for a mass political rally to be addressed by Hitler himself. The Doctor refused: he always maintained that the airships were for international transportation and commerce and not for political purposes. This stand got him into trouble with the Nazis then and would again when the *Hindenburg* was to take off at the start of the propaganda flight in March 1936.

The road and walkway into the airship plant crossed the landing field and many times when I was walking into the plant, Dr. Eckener would drive up in his Maybach car, stop, and give me a lift into the plant. Then, instead of immediately going up to his office, he would sit and talk about the mountains where I might have been hiking or skiing. He knew the mountains of Austria and Switzerland very well and always had suggestions as to which mountain I should climb the next time I was in a given area. These discussions about the mountains would sometimes go on for an hour or two before the Doctor would suddenly decide that he had better get to work.

By the summer of 1936 I was already well acquainted with "the old man," and he knew, I am sure, how I would react to certain conditions. It was obviously because of Dr. Eckener, since he had the final say, that I was aboard the *Hindenburg* on all its test flights—including the very first flight when I was the only one aboard who was not employed by the Luftschiffbau Zeppelin. There were no guests, no representatives of the Air Ministry, the United States Navy, or the news media. A month later, on the *Hindenburg*'s first South American flight, it was I whom he chose to confide in when he learned that Dr. Goebbels had declared him a "nonperson" and forbidden the German newspapers to mention his name or publish his photograph.

While at Lakehurst, the *Hindenburg* on October 9, 1936, made the so-called "Millionaires' Flight" over New England with 101 passengers, including many American industrialists, government officials, and senior officers of the U.S. Navy. Landing was at about dusk and by the time the ship was on the ground Dr. Eckener, I was sure, had had a very busy afternoon visiting with all his guests.

At the time I was on loan to the Germans to assist them at Lakehurst. When the *Hindenburg* landed I was in the vicinity and the Doctor, leaving the ship, saw me nearby. It seemed that everybody, particularly the news media and the photographers, wanted to talk with Dr. Eckener. He hurried toward me, leaving all the others behind, as if he had something very important to discuss. He began talking to me about something very inconsequential and I then knew what he was doing. He was a very tired man who wanted to get away from all the activity and needed a little peace and quiet.

We walked away from the crowd across the field to the road leading to the home of the commanding officer, who was then Commander Charles E. Rosendahl. The Doctor stopped talking as soon as we were out of earshot and the two of us had a quiet, leisurely, restful stroll across the field and on to the Rosendahls' home where I left him.

This was a pattern in the Doctor's behavior that I would see again and again in Germany. It was his way of getting away from the severe pressures of being a prominent public figure.

Chapter 6 *South American Flights in the* Graf Zeppelin

On June 3, 1934, George Lewis and Shortie Mills were back from South America aboard the *Graf*. It was understood that George and I would alternate trips aboard the *Graf*, and it would be my turn on the second flight of the year, scheduled to depart from Friedrichshafen on June ninth.

By now the transatlantic passenger flights of the *Graf* had settled into a routine. On the day of departure the passengers gathered, usually for dinner, at the Kurgarten Hotel. After dinner they proceeded by bus to the *Graf*'s hangar at the Zeppelin works, where they boarded the airship prior to undocking. The takeoff was scheduled for 8 P.M., though at times it might be delayed because the mail had not arrived on time, and on rare occasions, because of fog or thunderstorms in the Rhone valley in France.

The departure of the airship was fascinating both for the passengers and for those on the ground. With a ground crew of two hundred to three hundred men, depending on the weather conditions, the ship, moored to the traveling stub mast, was walked out of the hangar, made fast to trolleys fore and aft which ran on the docking rails. When clear of the hangar and mast, the ship was released from the trolleys and held on the ground by the ground crew. In an atmosphere of almost complete silence, final weigh-off was carried out by dropping water ballast to make the ship 900 to 2200 pounds light. At the command "Up Ship!" the *Graf*, released by the ground crew, silently floated upwards until at about 150 feet the engine telegraphs would be heard calling for the start of the first engine, then the second, third, fourth, and fifth engine. When the ship reached 300 feet with all engines idling, the engine telegraphs would again be heard signaling for higher revolutions on all engines. The lights in the control car and passenger area then gradually disappeared as the ship gained speed and altitude and moved off into the night with very little engine noise. All takeoffs were essentially the same: quiet, disciplined, and with the certainty of well-established routine.

Since the lifting capacity of the big rigids, and the payload, depended on their hydrogen content, and since the gas would blow off (through the automatic valves) if they climbed to high altitude, airships were flown as low as possible over land and sea, particularly at the start of a journey when heavily loaded with fuel. Hence the usual route for the *Graf Zeppelin* from Friedrichshafen to Rio led via Basel into France, flying by way of Besançon to Lyon and down the Rhone valley to Sainte Maries on the Mediterranean some fifty miles west of Marseilles. The Zeppelin *had* to proceed over French territory, and because of Gallic pride and the memory of the Great War, this posed a thorny political problem which became ever more acute with the

A view of the stern of the *Graf Zeppelin* in its hangar at night prior to the application of the swastikas and the German colors.

increasing excesses of the Hitler era. There was a constant suspicion on the part of the French that the airship was engaged in espionage and clandestine photography. In March 1929, the *Graf Zeppelin* made a cruise around the eastern Mediterranean. The director of the great Schneider-Creusot armament works thereupon alleged that the *Graf* had flown slowly just 650 feet above his plant: investigation showed that she had crossed at 2300 feet and at usual speed, and that all photographic equipment had been secured.[1]

In 1934 the *Graf* was restricted to a corridor twelve miles wide with terrain on both sides of the corridor considerably higher than the cruising altitude of the airship. Residents on both sides of the corridor were alerted and given a phone number in Paris, presumably that of the Air Ministry, to call if the airship passed over their area. Obviously, the *Graf* would not deviate from the corridor because of the definite possibility that if she did so, the Ger-

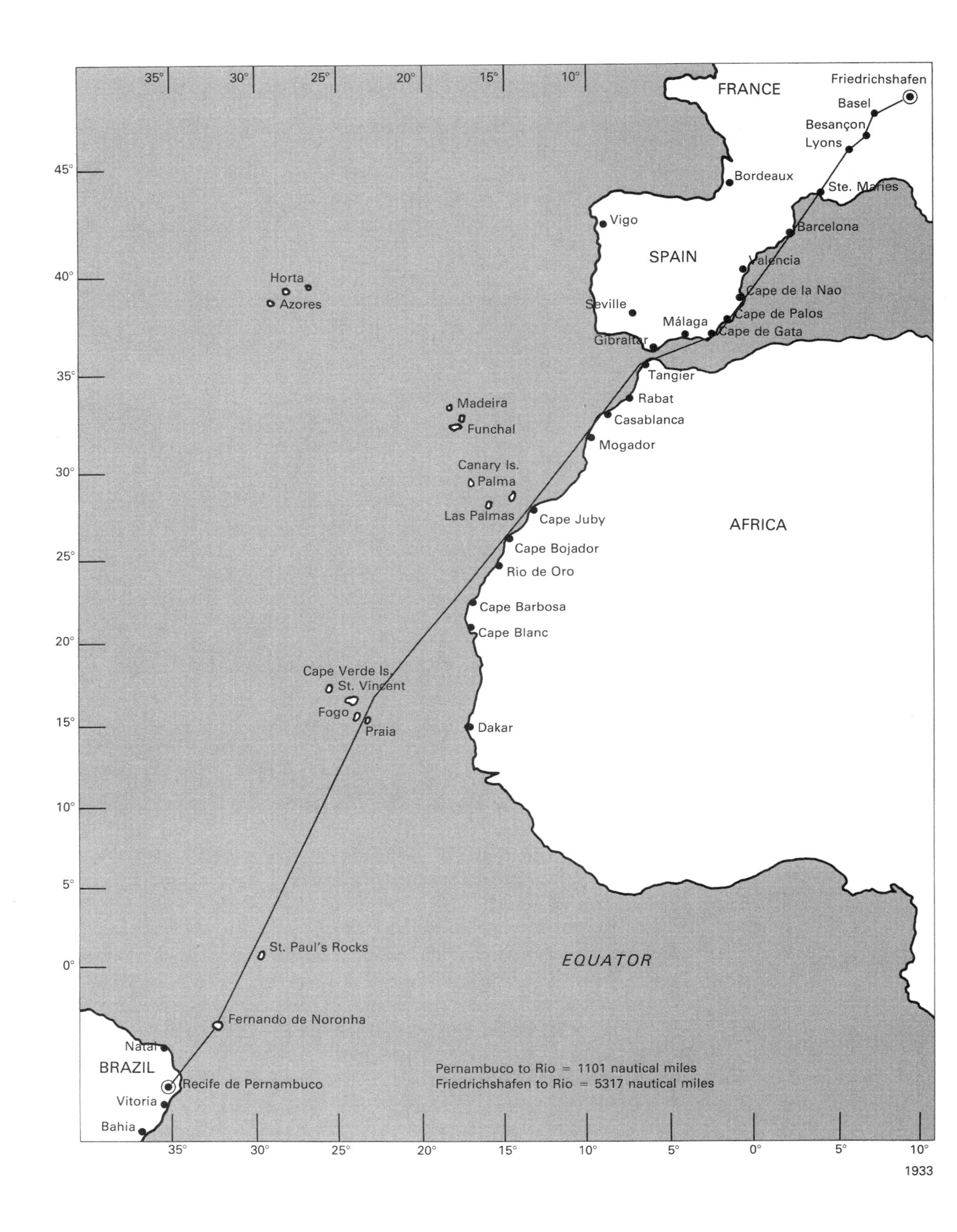

FRANCE
Friedrichshafen
Basel
Besançon
Lyons
Bordeaux
Ste. Maries
Vigo
Barcelona
SPAIN
Valencia
Cape de la Nao
Seville
Cape de Palos
Málaga
Cape de Gata
Gibraltar
Tangier
Rabat
Casablanca
Mogador
Horta
Azores
Madeira
Funchal
Canary Is.
Palma
Las Palmas
Cape Juby
AFRICA
Cape Bojador
Rio de Oro
Cape Barbosa
Cape Blanc
Cape Verde Is.
St. Vincent
Fogo
Praia
Dakar
St. Paul's Rocks
EQUATOR
Fernando de Noronha
Natal
BRAZIL
Recife de Pernambuco
Vitoria
Bahia
Pernambuco to Rio = 1101 nautical miles
Friedrichshafen to Rio = 5317 nautical miles
1933

The chart ON PAGE OPPOSITE shows the route of the *Graf Zeppelin* from Friedrichshafen to Recife de Pernambuco in South America. (From the collection of Harold G. Dick)

mans would lose the indispensable privilege of flying her over France on their way to South America.

On the morning after departure, the *Graf*, flying no higher than 650 feet, would follow the coast of Spain toward the Straits of Gibraltar. From the starboard windows of the lounge, passengers had a close-up aerial panorama of Barcelona, Valencia, Cartagena, and Malaga, while the navigational landmarks were Cape de la Nao, Cape de Palos, and Cape de Gata. At Gibraltar was the Rock, with its naval base and the ships of the Mediterranean fleet, symbolizing the power of Great Britain. As night fell, the airship proceeded southwest along the arid coast of French Morocco, with the cities of Tangier, Rabat, Casablanca, and Mogador going by on the port side. Next morning the *Graf* would be over Rio de Oro—the Spanish Sahara—where she would take her departure from Cape Bojador for the Cape Verde Islands. Here, with the islands in bright sunshine, the airship usually overflew the capital, Porto Praia, leaving to starboard the imposing 9281-foot volcanic peak of Pico da Coroa on Fogo Island.

Darkness again came with the airship well out over the South Atlantic, but by the early morning of the third day, and 1260 miles* further south, she was approaching the island of Fernando de Noronha with its soaring rock spire of the "Finger of God," 1089 feet tall. By evening the *Graf Zeppelin* was moored to the mast at Recife de Pernambuco, where the ship was refueled and any hydrogen valved en route was replaced. The average time from Friedrichshafen to Pernambuco, 4233 miles, was seventy hours. For the convenience of the passengers, the airship departed from the Pernambuco mast next morning and, at reduced power, flew down the coast of Brazil via Bahia and Vitoria, with passengers finding the scenery fascinating from the altitude of 650 feet. After a good night's sleep they arrived at Rio in the morning and debarked. The total distance, Friedrichshafen to Rio, was 5317 miles. The return flight averaged about ten hours longer because of the northeast trade winds' being ahead. Once again the ship came up the Rhone valley by night, to arrive in Friedrichshafen in the morning when the ground crew would be made up of workmen from the airship plant.

Flying through the Rhone valley presented problems because of the restrictions placed on us by the French government, and because of the weather conditions encountered there. When the valley was filled with thunderstorms, as was often the case, it was impossible to go around the storms because of the twelve-mile restriction. On the outward bound flight there was also the danger of valving gas during the storm, because a column of valved gas could act as a conductor to lead the lightning directly to the ship. With the mixture of hydrogen and air this could be disastrous. To avoid the possibility, when approaching a thunderstorm the ship was taken up through pressure height (the height or altitude at which all gas cells were fully

*The term "mile," here and throughout, refers to the nautical mile of 6080 feet.

A shipwreck on the African coast, one of the many wrecks visible from the airship. One could not help wondering how anyone could survive in this desolate area even if he did get ashore. (Photo by Harold G. Dick)

inflated to 100 percent capacity) by 150 feet and then the altitude was reduced by 150 feet, so there would be no danger of inflammable hydrogen's being valved in the area of the thunderstorm.

With lightning flashing on both sides and ahead of the ship, sometimes as close as 650 to 1000 feet (about one ship's length), it was almost impossible to tell the exact location of the lightning because of the dazzling brilliance of the flash. Under these circumstances each person in the control car called out his impression of the location of the flash and the commanding officer, making a determination on the apparent location of the flash, would immediately turn the ship in the opposite direction. Our route through the valley would be a zigzag determined both by the storm and by the boundaries of the twelve-mile corridor. Navigation was made even more difficult because the turning radius of a large airship is about one mile. With the corridor essentially blocked by a thunderstorm, the *Graf* would zigzag back and forth to avoid the heaviest lightning, often retracing her course. Finally when a less severe storm area was located, the ship would be taken through, frequently by reducing altitude and passing beneath the storm.

There were occasions when we believed the ship had been struck by lightning, but strange though it may seem to a layman, there was no particular danger (unless hydrogen was being valved) since the bonded metallic structure of the airship acted as a Faraday cage. The logic of reducing altitude in bad weather is that all vertical gusts as they approach the surface must become horizontal gusts and as such are easier to contend with than the vertical gusts. Since the Germans were so successful in the operation of their great rigid airships, it would appear that their logic was correct.

The "beautiful blue Mediterranean" was not always that way and some of the worst weather conditions were encountered over this body of water. For the passengers this could be even more interesting than the land masses passing below. On one flight proceeding towards Gibraltar with frontal conditions to the southeast, two water spouts (in effect, tornadoes over water) were seen, quite close together and both in contact with the sea surface. The airship held its course to the northwest of the water spouts and experienced no particular turbulence.

Normal cruising speed for the *Graf* was 72 MPH. On one flight to Pernambuco we encountered a front over the Mediterranean off Cape de Gata with heavy rain and a strong head wind. With head winds of 67 MPH for a period of twenty minutes, the ship made practically no progress. Some of the old hands aboard who had been flying airships for twenty years or more stated that in that entire period they had encountered head winds of such violence not more than three or four times.

Tail winds could be equally strong. Returning in the *Graf* from Pernambuco in December 1934, we made a brief stopover at Seville. When we were again over the Mediterranean, we encountered northwest winds of 45 to 56 MPH which gave the ship a ground speed as high as 106 knots. These winds, however, were spilling over the Sierra Nevadas and were extremely gusty. Under these conditions drift angles as high as 44 degrees were experienced (the highest ever recorded aboard the *Graf Zeppelin* was 49 degrees and that only once). Because of high aerodynamic gust loads, rudder angles under these conditions were restricted to 10 degrees.[2]

In this weather the seas were running very high through the Straits of Gibraltar. Freighters on an easterly course were taking green seas over the stern and the water spilling over the bows made it appear that the wake was *in front* of the ship! On the other hand, the Spanish fishing boats, with their high bows and sterns, were apparently designed for just such weather, and those boats we saw below were riding the seas and shipping no water.

The flight along the African coast offered the opportunity to view a remote area which at that time could not be seen by any other means. Ships could not get close to the shore, airplanes were not yet flying over this area, and there were no roads. From our usual cruising altitude of 650 feet we saw the rusty hulks of many ships—dozens of them—that had obviously been driven ashore. We could not help wondering what had happened to the crews and how they could have survived even if they had made it to shore. At one location along the coast there was an old abandoned fort that made one think of the novel *Beau Geste* and sagas of the French Foreign Legion. There were no signs of life around the old fort, only a trail leading in and out that seemed to come from nowhere and go nowhere. The fort stood alone, no vegetation, no dwellings, surrounded by the desert and an eerie silence.

Most of the coastal area merged with the sands of the Sahara. While one would have expected to see no life in this arid waste, we did see some nomads of the desert, in striped robes and turbans, on horseback, each man armed with a rifle. The *Graf Zeppelin* was a tempting target, and we knew they were shooting at us, for first we would see a puff of smoke, and then several seconds later we would hear the report of the rifle. To the best of our knowledge they never did hit the airship, but we certainly wondered how close they might have come.

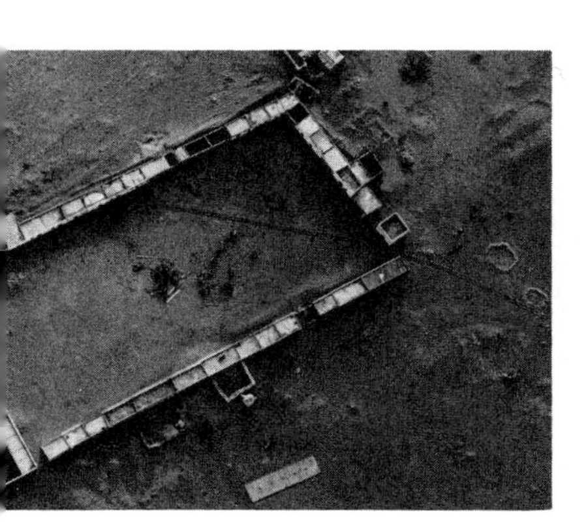

An abandoned fort on the African coast. There was no sign of life anywhere in the area, and the trail from the fort seemed to go nowhere. (Photo by Harold G. Dick)

The *Graf* crossed the equator on the long leg from the Cape Verdes to Fernando de Noronha, and there was an appropriate initiation ceremony and certificate for passengers who were "crossing the line" by air for the first time. One of my most prized possessions is the certificate given to me on the first crossing, though in my case it was to be the first of sixteen crossings, twelve in the *Graf Zeppelin* and four in the *Hindenburg*. The certificate was drawn up by Anton Eckener, Dr. Eckener's brother, and shows the South Atlantic bounded by South America and Africa, with the equator anchored to the African coast by a winch, the other end being made fast to a palm tree in South America, while Aeolus blasts away at the airship as it flies over the line. The certificate reads:

> We, Aeolus, Hippotes' son, friend of the Immortal Gods, rightful sovereign of the air, the weather, the winds and trades, monsoons and calms, have most

royally decided to give to the earth-born *Harold G. Dick* permission to make an aerial crossing of Our Equator on board the airship *Graf Zeppelin*. Given on board the *Graf Zeppelin* on the 12th day of June, 1934.

I have a number of signatures on my certificate that are quite noteworthy:

CAPTAIN FLEMMING was in command of the *Graf Zeppelin* on this particular flight. He was one of the World War I airship commanders.

CAPTAIN VON SCHILLER was a watch officer on this flight. He had acted as watch officer in five airships during World War I.

CAPTAIN PRUSS was a watch officer on this flight. He later had command of the *Hindenburg*.

CAPTAIN SAMMT, another watch officer, later was commander of the LZ 130, the *Graf Zeppelin II*.

CAPTAIN WITTEMANN, also a watch officer on this flight, was aboard on the last disastrous flight of the *Hindenburg* in a training capacity. This was in anticipation of his being one of the commanders of the LZ 130, then under construction, when it went into service. He also was one of the World War I airship commanders.

HANS LADWIG was navigator in Captain Wittemann's watch. This was the watch I also stood except when there was a shortage of hands, as when someone had to be left behind in Pernambuco because of sickness and I would then take over the watch on the elevator or rudder.

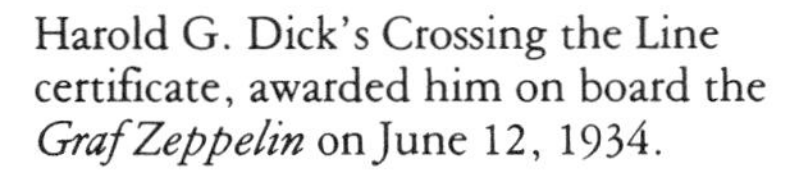

Harold G. Dick's Crossing the Line certificate, awarded him on board the *Graf Zeppelin* on June 12, 1934.

Wir Aeolus, des Hippotes Sohn,
ein Freund der unsterblichen Götter, rechtmäßiger Beherrscher
der Luft, des Wetters, der Winde und Passate, Monsune und Kalmen
haben allergnädigst geruht, de Staubgebornen
Harold G Dick
an Bord des Zeppelinluftschiffes „Graf Zeppelin"
Erlaubnis zum luftigen Überschreiten unseres Aequators zugeben
gegeben an Bord
des „Grafen Zeppelin"
am JUNE 12, 1934
Aeolus
m.p.

Fernando de Noronha, first landfall after crossing the South Atlantic. In 1934 this was a Brazilian penal colony; the surrounding waters were infested with huge sharks. The island is now a missile tracking station.

The *Graf Zeppelin* on the mast at Recife de Pernambuco. After the 5000-mile flight from Friedrichshafen, the ship was refueled and reserviced for the flight to and return from Rio de Janeiro. It was then again refueled, and hydrogen added as needed, for the return flight to Friedrichshafen.

Beautiful Fernando de Noronha was our first landmark after crossing the South Atlantic. The island is now a United States missile tracking station, but in the 1930s it was a Brazilian penal colony. It is completely green, hilly but not mountainous, with rock spires jutting up out of the sea around it. We never saw much of the inhabitants, who were few and far between. Most of the shoreline was steep and precipitous, but there were some sandy beaches inhabited solely by giant sharks. From the air we could see these huge creatures, perhaps twenty-five feet long, sunning themselves over the sandy bottom in water 25 to 50 feet deep. Obviously it was impossible for any of the prisoners to escape from this island!

At Pernambuco we were quartered in a building similar to a Quonset hut. Since Pernambuco is quite tropical, being only a few degrees south of the equator, and also only a few feet above sea level, all the bunks were completely enclosed with mosquito bars. At the canteen the standard drink was gin and quinine water—but nobody would take gin unless it was from a sealed bottle.

The *Graf Zeppelin* had one of her closest calls at Pernambuco on April 25, 1935, on her return flight from Rio. The airship arrived in the early morning but could not land immediately because of heavy rain and turbulence. After a two-hour delay, the captain decided to attempt a landing. During the approach a dense white cloud moving in from the sea was encountered. The *Graf* continued through this cloud which turned out to contain no rain and was of no great consequence. With the ship down to 300 feet and closer to the field, another cloud formation similar to the first was encountered. Because the first had been of no consequence, it was assumed this would be true also of the second: the airship, at this time about 1100 pounds light, was flown through.

The cloud in fact contained heavy tropical rain which within a few seconds added seven tons' weight to the ship. All available ballast, amounting to about five tons, was dropped but since the ship had little forward speed the extra heaviness could not be carried dynamically. The *Graf* settled to the ground about 2000 feet short of the field, completely carrying away the lower rudder, dragging the lower fin along the ground for 300 feet before the ship came to rest. The aft engine car bounced several times on the ground, and finally when the control car struck the ground, the floor of the navigation room caved in.

When the *Graf* came to rest, the chimney of one of the native huts was sticking up into the belly of the ship. A fire was going in the stove inside the hut. It was incredible that neither the hydrogen nor the Blau gas was ignited. One of the aft engine mechanics jumped from the car, ran into the hut, and with the pot of coffee that had been cooking extinguished the fire in the stove. Several palm trees were also sticking up into the ship.

Then, as rapidly as it had come, the rain stopped, the sun came out, and the *Graf* quickly became light. The mechanic barely made it back to his power car before the ship took off again. After she was again airborne, the lower rudder was seen lying on the ground about 600 feet from where the ship had come to rest. Since this was the first the *Graf*'s officers realized that they had lost the rudder, a man was immediately sent aft to see if the aft engine car was still with the ship. The landing was then made and well executed. The lower rudder was recovered and lashed in place on the ship, and after refueling the *Graf* was flown to Friedrichshafen with only the upper rudder operative. Damage to the ship was listed as follows:

> Lower rudder carried away. Longitudinal above the lower rudder broken. Outer edge of lower fin badly damaged. Aft engine car struts bent. Floor of navigation room damaged. Several rents in the outer cover. One gasoline tank punctured by a palm tree. Three breaks in the fuel lines. Minor structural damage, such as broken wires, trusses, channels, etc.

The *Graf* left Pernambuco on schedule and arrived in Friedrichshafen on schedule at 9 A.M. on a Tuesday. Thirty men worked almost day and night on the ship until Saturday when she left on schedule on her next flight to Rio.

The *Graf*'s commander on this flight was Captain von Schiller, who had joined the German Navy in 1912, had taken airship training at the very beginning of the war, had accumulated hundreds of hours flying in five naval airships in the years 1914 to 1918—and later in the LZ 126 and the *Graf Zeppelin*. Yet he had not followed the precepts laid down by Dr. Eckener.

The Doctor had been very successful in training operating personnel, first for the DELAG and later for the German Navy. He had also successfully commanded many airship flights, commencing prior to World War I. The basic premise for Dr. Eckener was that one could not *assume* that a situation was satisfactory; one had to *know* that it was satisfactory and then one could go on. If an assumption was incorrect, and a disaster occurred, the whole airship industry could be destroyed. It was therefore absolutely necessary to *know* that an operation would be successful before proceeding. This was his reasoning in handling any and all flight operations.

The incident at Pernambuco would not have occurred had the Doctor's reasoning been followed. The assumption concerning the second white cloud was incorrect and the *Graf* narrowly escaped a serious disaster. This was emphasized by the Doctor in no uncertain terms to the crew after the return to Friedrichshafen—emphasized to the extent that he could be heard up and down the hall outside his office with the door and transom tightly closed!

Chapter 7 *On-board Routine on Ocean Flights*

he Campo Affonso landing field used by the *Graf* in Rio was almost in the city close to the mountains and subject to fog. No mast was used, the ship being held on the ground by the ground crew while the passengers disembarked and the return passengers got aboard. This was far from an ideal landing place for a large airship, and for the *Hindenburg* a larger and better landing field with hangar was developed at Santa Cruz, about twenty-five miles south of Rio.

The ground crew was made up of Brazilian soldiers, who did not always have their minds on their work. During one landing, one of the ground crew at the aft engine car wanted to take a picture of the group at that station, so they all let go of the power car and turned to face the camera. The ship being light aft, the stern rose—and in short order the after car was about 100 feet above the ground, while the control car was still on the ground! All of us in the control car who were not required there were ordered aft as fast as possible. Ten men hurrying aft along the keel was approximately the same as one ton of ballast being moved aft, and this sufficed to bring the after power car back down on the ground. The men who did the hurrying were referred to as "galloping kilos."

The return flight over the South Atlantic was usually made at a higher altitude, normally 4000 to 5000 feet, and about two hundred miles off the African coast. At lower altitudes the northeast trades are augmented by an intense stationary low pressure area over the Sahara, but at higher altitudes the northeast trades diminish in intensity. At the higher altitudes, however, it was not uncommon to encounter the warm dry air from the Sahara, which at times would have as low a humidity as 6 percent. The presence of the air from the desert could be detected by a yellowing of the atmosphere as far as several hundred miles off the African coast. Today we would call it pollution, but it was the fine dust picked up by the winds and carried aloft. This dust, more like rouge, was deposited on everything aboard the airship. If this was too unpleasant for the passengers, altitude would be reduced to a level where they would be more comfortable.

Taking the ship into this warm dry air was almost like taking it into a front, for there was always some turbulence at the altitude where the warm air was encountered. Actually, rising from cool maritime air above the ocean into the warmer desert air above, the ship was ascending into an inversion and would become heavy on entering the warmer air. The line of demarcation was very sharp.

During the six round trips I made to South America in the *Graf* in 1934,[1] I had little leisure to enjoy the scenery as I was kept busy observing the operation of the ship and standing watches in the control car. I slept in a bunk in the crew's quarters, and ate in the officers' mess. I accumulated a total of 1196.36 hours of flying time, and covered 66,136 miles in the air.

Because the big airships were airborne literally for days on the transatlantic flights, the crew was organized in three watches on the same principle as in seagoing vessels. The full crew of the *Graf Zeppelin* was as follows:

Commanding officer (captain)	1
Watch officers (captains)	3
Navigators	3
Elevator men	3
Rudder men	3
Radio men	3
Sailmakers	3
Engineering officer	1
Keel engineers	2
Mechanics (for 5 engines)	15
Electricians	2
Cooks	2
Stewards	2
Total	43

The captain stood no watches but was always available. Control car watches were stood by watch officers and navigators, each watch being of four hours' duration, and each watch officer and navigator standing two watches every twenty-four hours. The first watch was 12.00–4.00 both morning and afternoon, the second was 4.00–8.00, the third was 8.00–12.00. These watches did not rotate, so that one watch officer always had the same watch. This was considered the best arrangement, particularly when flying over a standard route.

For the rest of the crew, including elevator and rudder men, the watches were of two hours' duration in the daytime and three hours at night. With three watches, the hours on duty for each were constantly changing. During the day each man was on standby (*Pikett-Wache*) for two hours before his next regular watch, and at night for the three hours before his next regular watch. Since there was little to be done at night, each man had six hours in which to sleep. There were also additional duties for all crew members:

The first radio man reported to the commanding officer on his equipment and similar concerns, but during his watch he was directly responsible to the watch officer, as were all the other radio men.

Elevator men, rudder men, and sailmakers (responsible for the gas cells, gas valves, outer cover, and ballast bags) had additional duties such as serving table in the officers' mess. There also were inspection duties as for example of the control lines; this was assigned to one man who made a thorough inspection twice a day.

Elevator or rudder men who were off duty ordinarily drew the weather maps, which were then turned over to the watch officer or commanding officer.

The first watch officer also served as the ship's executive officer. The second watch officer served as ship's purser. The third watch officer could be considered to be the chief navigator. This arrangement, however, was adjusted to the personalities and qualifications of the watch officers rather than being a part of the watch and station bill.

To ensure that I acquired complete knowledge of the operation of the *Graf Zeppelin,* I was assigned to Captain Wittemann's watch because his navigator, Hans Ladwig, spoke excellent English. I had no specific duties, under normal circumstances, except to be present during Wittemann's watch. Between watches, I was to familiarize myself with the complete operation of the ship, keel watch, gas cells, engines, and so on. Because knowledge of the operation and handling of the rudders and elevators is so important to understanding the feel and action of the ship, I would spend as much time as possible on these controls. Consequently when a regular rudder man or elevator man had to be left behind in Pernambuco, I would help out by standing that rudder man's or elevator man's watch. Additionally, if the elevator man were overloaded with his other duties, I would spell him off so he could get his other work done. Helping out was both good public relations and good experience for me—and there never seemed to be any question concerning my ability to handle either control.

Looking back, I find it surprising that I, an American, was so completely and promptly accepted by the crew of the *Graf Zeppelin*, and later of the *Hindenburg*, close-knit families that had been flying together for many years. Moreover, they never had to deal with foreigners, aside from our U.S. Navy airship officers who at times flew as observers. Even in 1934 quite a few crewmen were Nazi party members, at the same time being completely loyal to Dr. Eckener. Rudolf Sauter, the chief engineer of the *Hindenburg*, was a general in the Sturmabteilung (the Brown Shirts) and an ardent Nazi. He was also a good airshipman and very well qualified for his position; he was completely responsible for the maintenance and physical operation of the ship. It was said that when Dr. Eckener could not get what he wanted from the Air Ministry, Sauter, with his Brown Shirt connections, would go to Berlin and get it for him. Yet Sauter accepted me completely and I enjoyed several excursions with him and his friends. Only Heinrich Bauer, one of the watch officers of the *Hindenburg,* was a question mark. He would smile and remark that my mission was "open espionage," but I felt that his eyes were not laughing.

I learned about Dr. Eckener's procedure for navigating the *Graf* over water on the long legs to and from South America. The airship, with her relatively low speed, was much affected by the wind. Some means had to be developed to measure the wind's strength and direction, and the extent to which it caused the airship to be carried off course. Flying at low altitude as we did, the navigator every hour had the *Graf* turned 45 degrees off course, first to port with drift readings taken, then to starboard with drift readings taken. At night these readings were taken with the help of a 3,500,000-candlepower searchlight that was mounted under the belly of the ship abaft the control car. Wind triangles were then completed and we knew exactly the distance we were making good, how strong the wind was, and what was its direction.

One of our crossings of the South Atlantic was a fine example of the accuracy of this kind of dead reckoning navigation. For 1800 miles we had seen no landmark and navigation was entirely by this method. Our next landfall was to be Fernando de Noronha and we picked it up exactly on course and within minutes of the projected time. By comparison, celestial navigation was superfluous and radio bearings were grossly inaccurate.

Originally when contacting a ship at sea the policy was to request the ship's position and to correct the airship's position to correspond to that of the surface vessel. We soon found that this was not justified, as the seagoing craft seldom were where they thought they were. In fact we found that they usually were twenty-five to fifty miles from where they thought they were. So, when the positions did not coincide, we adopted the policy of suggesting to the ship's officers that perhaps they would like to correct their position to that of the airship. The navigators on the seagoing vessels were probably not too happy with us!

The altimeters we used were of the aneroid type, and the altitude recorded was a function of the barometric pressure. Although such an altimeter might read correctly for a short flight, it could be in error by as much as 500 feet if, for example, the airship moved from a high pressure area into a low pressure area. Such error could not be tolerated when making the landing maneuver, or at night or in thick weather. In order to correct the altimeter, a nonbarometric apparatus trade-named the Echolot was installed aboard the *Graf.* This consisted of a bouncing type of light indicator that was calibrated in meters of altitude, with zero being at the bottom of the scale. A gun, loaded with a blank cartridge, was fired downward through a sleeve in the control car when the bouncing light was at the bottom of the scale. When the echo returned from the surface, the bouncing light would vibrate and it was at this point that the altitude was read. The Echolot required a certain amount of skill to be read accurately, and was not used at night because it would disturb the passengers.

The Germans had therefore devised a very novel and accurate way of recalibrating the altimeter. Though they drink a lot of wine and beer, they also drink a lot of bottled mineral water—and all the empty water bottles were saved aboard the *Graf.* Previously a chart had been prepared plotting the time of fall against the height from which an empty mineral water bottle was dropped. To correct the aneroid altimeter a bottle would be dropped overboard, the time taken with a stopwatch, and the correct altitude immediately established. Bottles were dropped from the *Graf* in the South Atlantic (and from the *Hindenburg* in the North Atlantic), in the Rhone and

Rhine rivers, in the Bodensee at Friedrichshafen. The bottle altimeter had the distinct advantage of being usable at night—with the assistance of the searchlight—but only of course on over-water flights!

Weather was always an important concern for the *Graf Zeppelin*'s officers, who aimed to find favorable rather than unfavorable winds and to avoid the most severe turbulence as this not only would disturb the passengers but could overstrain the ship's structure, particularly in thunderstorms. Each portion of the route from Friedrichshafen to Rio had its own particular weather features. A common condition in the Rhone valley was the mistral, a strong north wind blowing down the valley. This phenomenon was caused by a high pressure area moving in from the northwest, centering over the British Isles, and then moving off to the northeast. The mistral was unpleasant for airship passengers because it produced a great deal of turbulence due to spillage over the mountains and sudden crosswinds.

The opposite phenomenon, the *Foehn* (as the Germans call it), was a warm south wind blowing in from the Mediterranean, and it occurred typically when sections of the low pressure areas centered over the Atlantic broke away and cut across the Bay of Biscay, over France, and into the Mediterranean. When this was accompanied by a high pressure area over Russia, the *Foehn,* which was usually not violent, increased to moderate strength.

Frontal conditions, which might accompany the motion of depressions with a counterclockwise rotation, moved from west to east across the North Atlantic. These fronts, when encountered, usually lay across the Mediterranean or to the westward of Gibraltar. Their chief characteristic was strong winds but not very great temperature changes. Particularly in the months of October, November, and December, there were high winds through the Rhone valley and across the Mediterranean to Gibraltar.

The doldrums, between the Cape Verde Islands and Fernando de Noronha, featured frontal conditions accompanied by tropical rains and squally air. The fronts lay in a northeast to southwest direction and when the front was encountered, the course of the ship was altered to port in order to pass through as quickly as possible. Temperature changes were often quite severe, the variation being from 9 to 18 degrees Fahrenheit. Not all frontal conditions involved clouds, rain, and thunderstorms. Some were clear, producing wind shifts and changes in temperature and humidity.

Such a front was encountered by the *Graf* over the South Atlantic when leaving the doldrums en route from Pernambuco to Friedrichshafen on June 18, 1934, on my first South American flight. Since there were absolutely no clouds, there was no visible indication of unstable air conditions, but an increased bumpiness of the air was noticed as the ship approached the front. This was even more noticeable on the controls. The ship was perhaps one ton light at this time.

On entering the front, the ship first pitched up and started to rise at about

400 feet per minute. This lasted only about ten seconds and then the ship pitched down and started to fall at 400 feet per minute. Altitude at this time was 650 feet. During this brief interval the temperature rose from 70 to 84 degrees F, the humidity dropped from 70 percent to 50 percent, the wind shifted from north by east 2–4 MPH to east northeast 29 MPH (equivalent to 57 degrees). Elevator angles of 10 degrees were being used while the ship was pitching at 8 to 10 degrees. The ship was also swung directly into the wind. After a quick check was made on wind velocity and direction at 650 feet, altitude was increased in hopes of finding more favorable conditions. Subsequent observations are shown below:

Altitude (FT)	*Temperature (F)* AIR	GAS	*Relative Humidity*	*Wind*	*Velocity*
650	84	77	50%	ENE	29 MPH
985	84	76	47%		
1310	82	75	44%		
1640	84	75	42%	E by S	24.5 MPH
1970	83	74	40%		
2300	84	74	38%		
2625	80	72	38%		
2950	79	69	38%	SE	17.9 MPH
3280	79	69	38%		
3609	78	68	38%		
3937	77	68	36%		
4265	75	68	35%		
4593	74	67	35%		
4922	72	66	35%	SE by S	17.9 MPH
5250	71	66	34.5%		

Rate of climb through pressure height 220 feet per minute. Cells full at 4593 feet: nos. 12, 13, 14, 15, 16; nos. 1, 2, 3 full at 4790 feet.

The southeast by south wind being more favorable, the ship was brought down to 4922 feet and continued on her course.

Thunderstorms were not uncommon on the coast of South America and were usually encountered between Pernambuco and Bahia. They occurred when cold air broke in from the south and the resulting temperature changes introduced severe turbulence. These thunderstorms could not be classified as dangerous, however, for they did not extend more than fifty miles out to sea and the ship's course could readily be altered to circumnavigate the center of the disturbance. Course was always altered to seaward. Common practice was to skirt the edge of the storm, studying its characteristics, and when a suitable opening was seen, the ship would be taken through.

Although all available weather maps were made up, it cannot be said that the *Graf* flew by the weather map. If conditions were too severe in the Rhone valley, the takeoff might be delayed, but this was primarily because of the

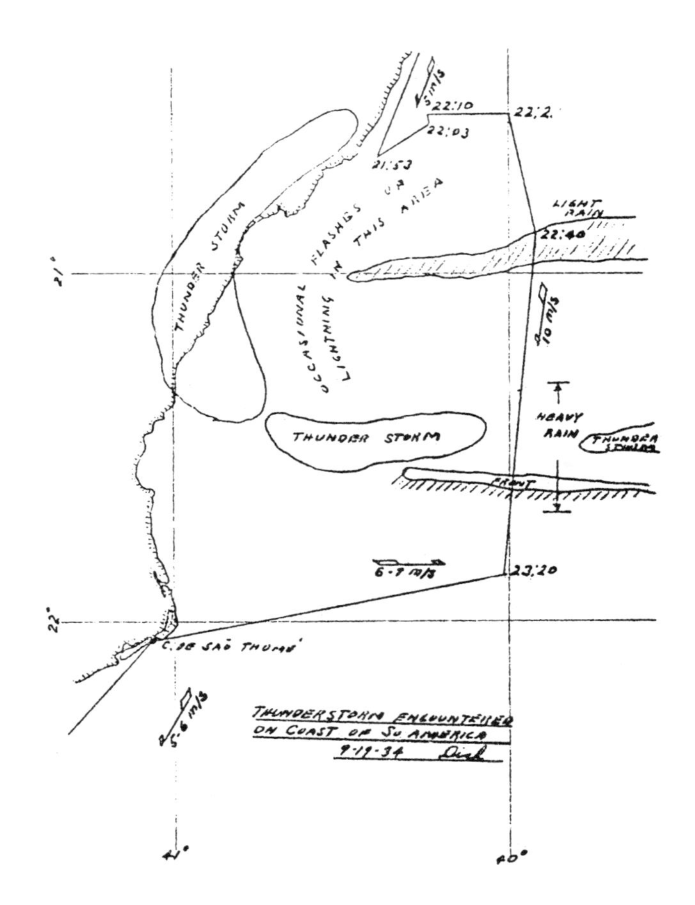

Chart of a thunderstorm encountered on the coast of South America in September 1934. (From the collection of Harold G. Dick)

restrictions imposed by the French. Takeoff might be delayed when the Rhone valley was completely fogged in, when there were strong head winds, or when the valley was full of thunderstorms. The twelve-mile passage way was insufficient for maneuvering around severe disturbances. It may be said that the weather reports obtained aboard the *Graf* served primarily to forewarn the officers in case of extraordinary circumstances. Normal weather conditions as outlined here were judged by the officers when encountered, and the ship's course might be altered accordingly.

Chapter 8 *Airship Flight Procedures as Taught by Dr. Eckener*

he most interesting and valuable information made available to me was that involving the operation and handling of a big rigid airship. This was more an art than a science, refined to a high degree by the Germans under Dr. Eckener. Their incomparable record included thousands of journeys safely completed. Operating an airship is more complicated than operating an airplane. The airplane is sustained in flight by the dynamic forces of air acting on a wing which is driven through the atmosphere by the thrust of the engines. Thus, the airplane may be called an *aerodyne.* Distinctly different, the airship (and the balloon, since the airship is merely a powered balloon) is sustained in the air when motionless by a medium lighter than air. Thus, the airship may be called an *aerostat.* For serious operations, there are only two gases to consider, hydrogen and helium. Since nonflammable helium was then found only in the United States, and its export was forbidden by law, the Germans had to make do with inflammable hydrogen, which had cost them many lives in wartime even before the *Hindenburg* disaster.

A thousand cubic feet of pure hydrogen under standard conditions[1] weighs 5.61 pounds. The same volume of air weighs 80.72 pounds, the *net* lift being therefore 75.11 pounds per 1000 cubic feet. *There is no means whatever* of exceeding this static lifting force, and indeed in practice it is invariably less owing to contamination of the gas with air. The Germans used a net figure of 72 pounds per 1000 cubic feet in their lift calculations.

The gross lift of the *Graf Zeppelin* at Friedrichshafen (1300 feet above sea level) with 3,250,000 cubic feet of hydrogen was 210,000 pounds. The weight empty was 150,000 pounds, and the useful lift was 60,000 pounds. This was available for the carriage of passengers with their baggage and supplies; the crew and their effects; cargo and freight; water ballast, stores, and spare parts; and gasoline if this were carried in addition to the essentially weightless Blau gas. Any change, however, in the barometric pressure, humidity, air or gas temperature, or gas purity would alter the gross lift and affect the useful lift. Generally speaking, the lift was greater with a high air density, with cold temperatures and high barometric pressures, and less with high temperatures and lower barometric pressures. Therefore, a rigid airship could lift a greater load of fuel, passengers, and cargo in winter than in summer. Obviously this was a handicap to the *Graf*'s operating in the tropics. Humidity had a relatively slight effect, though a high air humidity somewhat decreased the lift. In addition, the hydrogen was constantly diffusing outward and the air diffusing inward, even with "gastight" material such as goldbeater's skin, and this contamination obviously increased the weight of the gas and decreased the useful lift. During the few days' layover in Friedrichshafen between the flights to South America, and again at Pernambuco on arrival from Europe and before the flight home, the gas cells were topped up to 100 percent fullness with fresh hydrogen, maintaining the purity at a high figure.

Temperature, barometric pressure, humidity, and gas purity would be considered by the airship captain in loading his craft in the hangar, while great care was taken that the weights were distributed along the keel in such a way that the center of gravity was exactly under the center of lift, and the airship was neither bow nor stern heavy. On being walked out, the *Graf* was "weighed off" and in equilibrium, with the weights and lift of the gas evenly balanced. Several hundred pounds of water ballast was then dropped to give the ship static lift at takeoff.

Thereafter the static lift of the ship would vary constantly. If she were 100 percent full of gas, as was the German practice on takeoff, she lost lift as she ascended while the expanding hydrogen spilled from the automatic valves. Thus, an airship usually was heavy during the early part of a flight, but became statically lighter as the journey continued and she burned fuel. Being a powered aircraft, however, the big rigid airship, like the airplane, could develop dynamic lift, with the hull when inclined upward or downward acting like a crude airfoil.

The *Graf* with all engines on cruising power could carry an overload of as much as eight tons by flying 5 degrees nose up, or a corresponding degree of lightness could be compensated for by flying 5 degrees nose down. (In an emergency the *Graf* could carry twelve tons at 12 degrees with all engines on full power, but this was never necessary as far as I know.) Beyond two degrees the *Graf*'s operators considered that the drag would be unacceptable and the passengers uncomfortable, and the ship would have to be brought into static equilibrium—by dropping water ballast if she were heavy (the *Graf* had the option of burning gasoline instead of Blau gas to make her light), or valving hydrogen if she were light. In addition, the static condition of the ship varied when the gas was not of the same temperature as the air. "Superheating," usually from the heat of the sun warming the gas, made the ship light as the hydrogen expanded, while "supercooling" after dark caused the gas to contract, making the ship heavy.

My previous experience in the United States in 1933–34 flying in balloons and later the Goodyear blimps (mostly in the *Defender,* the 178,000-cubic-foot flagship of the Goodyear fleet) had given me considerable practical knowledge of aerostatics, and enabled me to take my place with the crew of the *Graf Zeppelin* as something more than a raw beginner. However, there was much more to learn from the German experts about flying a big rigid, and I compiled voluminous notes on the handling of the *Graf Zeppelin* in various conditions:

> It is preferable to fly ship as nearly in equilibrium as possible. If ship is more than one ton heavy gasoline is generally burned to reduce heaviness. Generally, ship is flown on the light side more than heavy. On the average, ship is flown within ±

1 degree of pitch. Angles of pitch indicate the following amounts of heaviness at 71.6 MPH:

1 degree	2 tons (4400 lb)
2 degrees	2 tons plus 1.5 ton
3 degrees	2 tons plus 1.5 ton plus 1 ton

Altitude is checked with the Echolot every morning and evening, and generally during the day, either with the Echolot or by dropping bottles. Immediately after passing through front altimeter is generally checked. If in flying any sort of distance the accuracy of the altimeter is doubted, it is immediately checked. As a rule, in order not to disturb passengers too much, it is preferred to determine the correct altitude by dropping a bottle rather than by gun sounding. Altimeter *always* checked immediately before landing.

Great importance is attached to the position of elevator man, it being essential to have a man at that position who has the "feel" of the ship and one who reacts quickly. By preventing the ship from getting a start either up or down before being checked by the elevators very few sudden ascents or descents result.

In flying through fronts, a light spot is picked in the clouds and as a rule the cloud bank is flown *under* at this light point, for at lower altitudes air is not so turbulent as higher up, due to the flattening out of vertical gusts against the surface.

The usual flying altitude above the surface is from 575 to 820 feet, unless more favorable winds can be found at higher levels. If in flying at greater altitudes, however, and rough weather is encountered, the lower level is invariably resorted to and an effort is made to fly under the disturbance rather than directly through or over it.

Ship's officers never hesitate to turn back and look for a softer spot if they find ship meeting conditions that are too tough. When flying into approaching squall line or thunderstorm the nose is kept down, even if the ship is heavy, for the first vertical gust is always up.

Mass: Since the dead weight of the ship is 150,000 lb (flight figure) (146,000 lb Design Department figure) and the total of all other loads is about 55,100 to 66,100 lb, the mass of the ship for all approximate cases may be taken as 220,000 lb or 100 tons. [NOTE: While the ship, when weighed off—with lift and loads evenly balanced—may appear weightless, this figure indicates the enormous mass and kinetic energy of a large rigid airship like the *Graf* when once set in motion by her engines.]

Emergency ballast: [This in effect refers to the "breeches" of 660 lb capacity each, of which four are carried forward at Frame 195, and four in the tail at Frame 20. A pull on the toggle in the control car instantly dumps the entire 660 lb of water.] This is classified as "landing ballast" and when no intermediate landings are being made, such as on a flight [from Friedrichshafen] to Pernambuco, only 2640 lb are carried, that is, 1320 lb forward and 1320 lb aft. When intermediate landings are to be made and takeoff conditions are at all uncertain, more of the emergency bags are filled and the ship may take off with 3960 to 4620 lb of emergency ballast. Capacity is 5280 lb.

Trim ballast: This is made up of the remaining water ballast aboard (carried along the keel and controlled from the control car) less the drinking water and wash water. Total capacity is 17,600 lb.

Takeoff: For long flights the inflation is carried to as near 100% as possible, but temperature changes may give the ship a pressure height of a few feet (100 to 165). The takeoff is usually made with the ship 660 to 1320 lb light, and the altitude reached before the ship is in equilibrium is 33 feet for each 220 lb of lightness. For each 330 feet which the ship is forced through pressure height, it becomes 1 ton heavy (through loss of hydrogen through the automatic valves) and all engines burn gasoline until a state of equilibrium is again reached.

Automatic valves: Each cell is equipped with an automatic valve with the exception of the after end cell which is connected to the adjacent cell. When flying the ship through pressure height, 1% of gas is lost for each 330 feet. Bow and stern cells, which have a lower pressure height since there are no fuel gas cells in these bays, lose 9890 cu ft for each 330 ft ascent, and the center cells lose 30,366 cu ft per 330 ft. (The total volume of the ship is 114,000 cubic meters[2] or 4,025,340 cu ft). When the pressure height of the center cells has been increased due to the use of the fuel gas, and altitude is being increased, the inflation sleeves are connected from the bow and stern cells, and instead of valving from these cells the gas flows to the other cells.

Maneuvering valves: Six center cells (nos. 4, 7, 8, 9, 10, 13) have their valves connected to the hand wheel in the car (so they can be opened simultaneously for a given period of time) and this is used in making weigh-offs. The adjacent cells are controlled individually for trim purposes. Each valve has a capacity near sea level of 1765 cu ft per minute (approximately 100 lb/min) at about 70–80% inflation.

Fuel consumption: With 5 engines at cruising, the gasoline consumed is 926 lb/hr, and the fuel gas consumed is 8827 cu ft/hr, which will make the ship 110 lb/hr lighter and will increase the pressure height, if the ship is at pressure height, by 72 feet per hour (because the Blau gas, with a specific gravity of 1.06, is actually slightly heavier than air). (Though an added note, attributed to Sammt, says "actual flight consumption is greater due to gusts, turbulence of air, etc.")

Trim ballast: This is located in 8 bags (distributed along the keel) giving a capacity of 17,640 lb. Each bag is equipped with a dump valve allowing water to be discharged at a rate of 2.2 lb/sec or 132 lb/min.

Water capacity: The total water capacity may be broken down as follows:

Emergency = 8 × 660 = 5280 lb
Trim = 8 bags = 17,600 lb
Drinking = 4 × 880 = 3520 lb
Total = 26,400 lb.

Wash water arrangement: The ballast arrangement on the *Graf* is such that all bags may be classified as for either clean or dirty water, there being an equal capacity for both. When ballasting the ship before takeoff the water is first run into the fresh water bags, and if additional ballast is required the water is then run into the dirty water bags, but of course this cannot be used for washing or other purposes.

During flight the fresh water is pumped to the fresh water bag at Frame 165, from where it is pumped to the header tank at 170, which provides the head for the wash basins. Used wash water is collected and pumped into the bags designated as dirty water bags.

The ballast system is interconnected by two lines, the one on the port side being for fresh water only, and the line on the starboard side being for dirty water only. These are cross connected at Frames 65, 140, and 165, in addition to the fittings required to the emergency bags which make it possible to fill all emergency bags before takeoff if so desired.

The general set-up is that the bags on the port side of the ship are for fresh water, and the bags on the starboard side are for dirty water. Toggles in the control car are either black or white to designate dirty or fresh water.

The bags are located as shown below:

Port		Frame	Stbd.	
1320	fresh	20	1320	dirty
		30	2200	dirty
2200	fresh	45		
2200	dirty	65		
2200	fresh	80		
		140	2200	dirty
		155	2200	fresh
2200	fresh	165	2200	dirty
		170	header tank—fresh	
1320	fresh	195	1320	dirty
Total fresh = 11,440 lb			Total dirty = 11,440 lb	
Cross connected = 9240 lb				

Some of the most obscure and unexpected problems were the lot of the elevator man when flying the *Graf Zeppelin* either heavy or light. The ship was almost always operated in trim. It was the chief engineer's job to keep it so, usually by pumping fresh or dirty water forward or aft. Paradoxically, while the ship in this condition required little elevator angle to counteract pitch oscillations, this was *not* the most stable condition in which the ship could be flown.

When flown statically heavy, the ship always tends to nose up and climb, while when light, she puts her nose down and tries to descend. The reason is that with the hull acting as an airfoil, and meeting the air stream at an angle, *the center of pressure on the hull moves forward,* creating a pitching moment either upward or downward. To the elevator man, however, it *appears* that the center of gravity has shifted, towards the stern when flying heavy and nose up, and towards the bow when flying light and nose down.

The *heavy ship in trim* will always tend to nose up and climb, and requires *down* elevator, the amount depending on the amount of heaviness. In bumpy weather, the amount of up elevator is less than the amount of down elevator, since the ship always tries to come back up by the nose. A ship half a degree heavy and in trim is the most stable of all. In bumpy

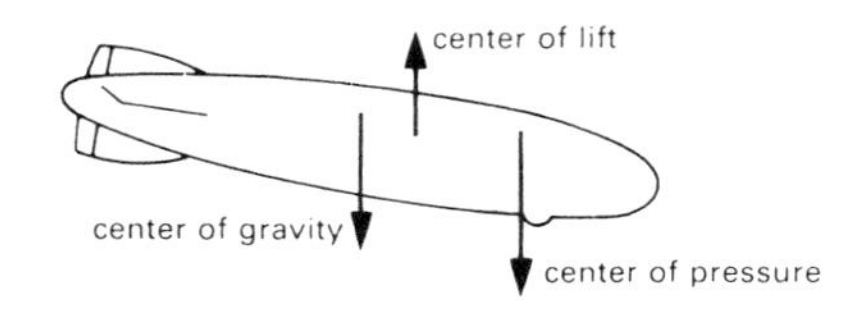

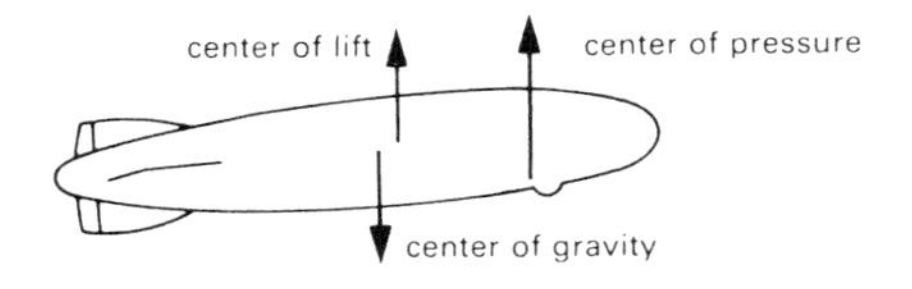

Forces acting on a rigid airship in flight. TOP: The light ship tends to rise statically. Nose goes down because of center of pressure moving forward, and is counteracted by *up* elevator.

BOTTOM: The heavy ship tends to descend statically. Nose goes up because of center of pressure moving forward, and is counteracted by *down* elevator.

weather, the upward gusts will appear more severe, since the ship will be entering them up by the nose.

For the *light ship in trim,* the stability is not as good as for a heavy ship. The ship will normally fly nose down and will always require up elevator. Altitude cannot be closely held and the elevator angle is constantly varying to keep the ship from rising and falling. Should the center of gravity shift forward or aft, more or less elevator might be required to keep a heavy or light ship at uniform altitude.

Though the *Graf* might be flown dynamically when heavy or light, she again became an aerostat when she came in to land with little or no way on. The dynamic lift decreased with the square of the velocity, and at low speed would be negligible. Thus, it was vital that before a landing was attempted the ship would be "weighed off" to determine if she were heavy or light, and then be brought into equilibrium. With the engines throttled back, the watch officer on duty would watch the variometer to see if the ship were rising or falling. A chart showed how heavy or light she might be for any rate of ascent or descent. If the airship with engines idling sank at the rate of 300 feet per minute, she was about 1100 pounds heavy and a corresponding amount of water ballast would have to be dropped. If she rose at the same rate, a corresponding amount of hydrogen would have to be released (a minute and a half with the six midships maneuvering valves open and connected to the wheel).

Before attempting a landing, contact was made by radio or flag signals with the ground crew to determine the ground temperature and wind conditions. For a normal calm weather landing the ship was trimmed very slightly nose down, as this gave a better gliding angle and the ship almost flew herself down. The airship displayed a white flag with a red St. Andrew's Cross (the international signal for "V" or "Victor") as a landing signal, and a smoky fire was started on the ground to show the wind direction. The ship then made a long approach with a rate of fall of 100 feet per minute, and the lines were dropped when she was over the landing flag. When conditions were unusual, as in gusty and bumpy weather, the *Graf* was weighed off a little light, and the approach had to be fast and preferably long and low.

When the airship was over the field the engines had to be reversed for some time to stop her, and this also helped to get the nose down. Yaw lines dropped from the ship's nose were drawn out to port and starboard by thirty men each, while twenty more on each side pulled the ship down with spider lines (so called because twenty short lines radiated like the legs of a spider from a block). When the airship reached the ground, fifty men held the control car rails and twenty held those of the after car. With thirty men in reserve, the ground crew totaled two hundred men.

One more factor to consider when flying in the tropics was the tremendous area of fabric. In a large rigid airship such as the *Graf,* the outer cover and the gas cells were all made of cotton. The outer cover was waterproofed with five coats of dope, the last two containing aluminum powder, while the gas cells were lightly coated with soft wax outside and varnished inside. Nonetheless, the fabric absorbed considerable moisture, adding to the weight empty of the *Graf* and decreasing the useful load. An increase of relative humidity from 60 to 80 percent increased the weight by 2200 pounds, and a further 10 percent increase to 90 percent—not uncommon in Pernambuco—increased the weight by another 1760 pounds.

I include some tables showing the actual flight loads for the Friedrichshafen to Pernambuco leg of my first flight to South America in the *Graf Zeppelin* from June 9 through June 12, 1934. The sheet headed "Ballast for

Actual table showing water ballast on board, by location, at takeoff and before landing on the Friedrichshafen to Pernambuco leg of Dick's first flight to South America on the *Graf Zeppelin*, June 9–12, 1934. (From the collection of Harold G. Dick)

Wasserballast an Bord.

BENZ : 8040
OIL : 770 + 150
GAS : 19,000 m³

Fahrt Nr. 366

Datum: am, vom 9: 6/9/34 bis

von FRIEDRICHSHAFEN nach PERNAMBUCO

Ort	Nach Start	Vor Landung
Ring 20	H. B. B. H. St. B. 600 kg	2 H. B. B. H. St. B. 600 kg
Ring 35	—	500
Ring 45	550	—
Ring 65	—	500
Ring 80	1000	450
Reserve-Kühlwasser	2 Sack à kg [200]	2 Sack à kg [200]
Ring 140	850	800
Ring 155	—	400
Ring 160	1 Faß B. B. 2 Faß St. B. [1200]	300 Faß B. B. 150 Faß St. B. [450]
Ring 165	kg B. B. kg St. B. 700	300 kg B. B. 450 kg St. B. 750
Wasch-Wasser 175	[200]	[220]
Küche 190,5	[75]	[75]
Ring 195	2 H. B. B. 1 H. St. B. 900	2 H. B. B. 1 H. St. B. 900
	[1675] 4600	[945] 4900
Insgesamt	6275 kg	5845 kg

		Bemerkungen: 430 kg USED FOR DRINKING AND COOKING + TOILETS
Res.-Kühlwasser	200 kg	
Waschwasser / Trinkwasser	1475 / 3100	
Fahrballast		2500
Landeballast	1500	2100

Abgabe beim Start .... kg
" während der Fahrt .... "
" beim Abwiegen .... "
" beim Landen .... "

Bemerkung: [ ] nicht sofort ziehbarer Ballast.

Flight No. 366'' shows all loads including ballast, inflation and lift calculations, fuel loaded and consumed, and flight data. The ''Wasserballast an Bord'' sheet reproduced here shows water ballast on board by location, at takeoff and before landing. The third sheet shows the location of all loads throughout the ship in relation to the numbered transverse frames.

Ballast for Flight No. 366 *Friedrichshafen to Pernambuco*	*Loading*
Passengers (14 + 3)	1360 kg
Baggage for passengers	400
Baggage for crew	400
Provisions	1200
Freight	220
Mail	170
Drinking water	1475
Additional drinking water	500
Water ballast	2500
Absorbed moisture	300
Lift	800
Total	9325 kg
Engine personnel (20)	1600 kg
Deck force (23)	1840
Gasoline for 20 hrs (420 kg/hr)	8040
Oil	920
Kerosene	100
Spare parts	500
Reserve radiator water	200
Landing ballast	2100
Miscellaneous	1600
Fuel gas for 76 hrs (19,000 cbm)	1100 250 cbm/hr
Total	18,000 kg
Total Loads	27,325 kg at 80.7% H_2 fullness
Dead weight	68,000
Total weight	95,325 kg
Required lifting gas	92,000 cbm
Bow and stern cells	28,115 at 100% at 28,105 cbm
Center cells	63,895 at 74.5% at 85,895
Total	92,000 cbm = 80.7% of 114,000
Remaining volume	22,000 cbm = 19.3%
Fuel gas for 76 hrs. (729/21)	21,200 cbm = 18.7%
Volume to reach ceiling	800 cbm = 0.6%

Gas and fuel consumption

Fuel gas	14,300 cbm
Gasoline	3,040 kg
Oil	487 kg
Lifting gas	600 cbm

Flight data

Takeoff	6-9-34	19:10 GMT	
Landing	6-12-34	09:47 GMT	
Elapsed time			62 hrs 37 min
Distance			7843 km

Lifting constant	$v = 80.7\%$
$b = 729/m/m$	$b = 729\ m/m$
$tl = 18.0$	$t/L = 18.0$
$s = 0.094 = 1.036$	$t/G = 21.0$
	$s = 1.05$
	$R.h. = 70.0\%$

Flight No. 366
Friedrichshafen to Pernambuco — *Loading*

Ring	Fuel Gas	Miscellaneous Reserve Parts	Oil	Gasoline	Water	Baggage	Mail	Provisions	Passengers
–2									
10									
20					600	80			
35									
50					550	320			
65				1000					
80		1600	500	1000	1000			700	
95						560	220		
110	212,000			4000	200		170		
125	cbm	500	420			560			
140				1000	850			500	
155				1040	1200	400			
170					900	160			
185					75			500	1360
195					900	1360			
207.5									
215									
kg	1100	2100	920	8040	6275	3440	390	1700	1360

Note: 430 kg water used during flight for cooking and drinking.

Attention is called to the remarkably small expenditure of lifting hydrogen and water ballast on this long flight, covering 4232 miles in 62 hours 37 minutes at an average speed of 77.8 MPH. With 3,248,520 cubic feet of hydrogen in the gas cells at takeoff, only 21,000 cubic feet was valved off en route! Moreover, the water ballast sheets show that only 946 pounds of water was expended, and this not to lighten the ship, but for drinking, cooking, and toilets.

It had not always been so. Kapitänleutnant Ludwig Bockholt, commanding the German Navy's L 59 in the Africa flight in November 1917, had been the first to fly a rigid airship in the tropics, and his report on operational problems was discouraging. With a gas volume of 2,418,700 cubic feet of hydrogen, gross lift of 175,300 pounds, and useful lift of 114,400 pounds, L 59 at takeoff was loaded with 47,800 pounds of gasoline, 3363 pounds of oil, 20,200 pounds of water ballast, and 35,800 pounds of cargo, and had twenty-two men aboard weighing 3880 pounds. During the four-day flight she burned 25,964 pounds of fuel and 1107 pounds of oil, while 930,000 cubic feet of gas was valved off. On an average, L 59 was lightened by the burning of 6491 pounds of fuel and 247 pounds of oil each day. In contrast to the *Graf*, whose five Maybach VL-2 engines gave a total of 2750 HP for dynamic lift, L 59 had five smaller Maybachs developing a total of only 1200 HP, and the forward engine failed permanently on the second day over the Egyptian desert. Thereafter she was flown much of the time on only three engines, as each one was shut down in rotation for two hours' maintenance in every eight hours. With more engine power L 59 could have carried a larger load dynamically. Bockholt noted: "On the long distance flight, 5500 pounds heavy, flying at 6 degrees with four engines at 1260 RPM, ship held up well even at 10,000 feet with enormous decrease in speed and almost complete failure to respond to the rudders. At 7 degrees fell slowly. From my experience the type should not be flown continuously at more than 3 degrees."[3]

On the second day, proceeding south across the Egyptian desert at 3000 to 3300 feet, Bockholt found an air temperature of 50 degrees F, dropping to 46 degrees at noon at the Farafrah Oasis. With superheat of 26 degrees F expanding the gas, the ship was very light, being completely dried out in the bright sun and very dry air, and flew nose down in very bumpy air with an excess load aft of 1650 pounds to counteract the nose-down tendency in the light condition. Trouble developed that night as L 59 reached the Nile valley, and entered the area of the northeast monsoon. The air temperature after dark rose to 72 degrees F, and at 3 A.M. to 77 degrees, while the gas was supercooled by 9 degrees F. Much gas having been lost during the day through the automatic valves, Bockholt at sunset had dropped 4400 pounds of ballast and expected that the ship would fly heavy at 4 degrees nose up on four engines. In the warm air the engines overheated and lost power, and at 3 A.M. the ship stalled, fell from 3100 to 1300 feet, and was checked only by stopping all engines and dropping a further 6200 pounds of water and

The "Africa ship," L 59, which in 1917 flew nonstop from Bulgaria to Khartoum and return, a total distance of 4200 miles, carrying 15 tons of cargo for the German troops still fighting in East Africa.

The L 57, a sister ship to the L 59. L 57, 743 feet long, was built to fly to Africa with ammunition and medical supplies.

cargo. At about this time L 59 was recalled by radio from Germany and turned back, having reached latitude 16°30′ N, not far from Khartoum.

Rising to 10,000 feet for the return journey, Bockholt was lucky to find a tail wind for the flight home, but valved considerable hydrogen in the process. On the fourth night, flying at 1300 feet above the mountains of Asia Minor, at an up angle of 5 to 6 degrees with four engines running at full power, the ship stalled and fell abruptly in gusty winds off the mountains; 6600 pounds of ballast and cargo had to be dropped. "To fly steadily 4 degrees heavy at night can easily be catastrophic," noted Bockholt, "with sudden temperature changes in the Sudan . . . particularly if the engines fail from overheating with warm outside temperatures. On the first and third evenings, when the ship was made properly light after cooling, she flew well all night, while on the second and fourth evenings, saving ballast and flying at 4 degrees, a lot of ballast and some of the ammunition had to be dropped. After weighing off for the original flight altitude, the ship should have 6600 pounds or 4 percent of her lift to drop for each night to take care of cooling effect."[4]

So there it was—for each night of flight in the tropics, an airship the size of the *Graf* would have to have 7520 pounds of ballast to drop, carried at the expense of fuel, useful load, and payload—a hopeless proposition for a commercial aircraft. Dr. Eckener, as advisor to Fregattenkapitän Peter Strasser, the German Navy's Leader of Airships, naturally had complete access to Bockholt's war diary and reports, and knew he had to find some solution to this problem.[5]

The Blau gas fuel was the answer; because it weighed only slightly more than air, its consumption did not lighten the ship and force the valving of hydrogen by day, with resulting cooling and heaviness after dark. On my first flight to South America, for which loading sheets are presented above, 748,572 cubic feet of fuel gas had been piped aboard, sufficient for 76 hours of cruising flight. During the 62½-hour flight, 504,933 cubic feet had been consumed, indicating that the light Blau gas had been used for a total of roughly 56 hours. Of the total of 17,728 pounds of gasoline loaded, sufficient for 20 hours of cruising flight, 6700 pounds had been consumed indicating that the liquid fuel had been burned for a total of about 6½ hours when the ship was heavy. During the entire flight from Friedrichshafen to Pernambuco, the *Graf Zeppelin* was lightened only by 2440 pounds (fuel gas burned) + 6700 pounds (gasoline burned) + 1073 pounds (oil consumed) + 948 pounds of water used for drinking, cooking, and toilets—a total of 11,161 pounds. The 21,000 cubic feet of hydrogen valved during the voyage compensated for about 1500 pounds of the lightness, with more hydrogen being valved during the weigh-off before landing. What a contrast with the experience of L 59 over the African continent, which indicated that the *Graf* would have to release 7520 pounds of ballast each night to compensate for a loss of 105,000 cubic feet of hydrogen each day!

Further, the maritime tropical environment of the over-water South American flights did not present the temperature extremes of the Egyptian desert. Lastly, if the *Graf* became suddenly heavy, as in a tropical rain squall, she could carry nearly three times the load dynamically that L 59 could support with her four 240 HP Maybachs.

The *Graf*'s operations contrasted also with those of the U.S. Navy's rigid airships. These had the advantage of being inflated with nonflammable helium, but since this was heavier than hydrogen (11.14 pounds per 1000 cubic feet in the pure state), their gross lift was only 93 percent of that of the hydrogen-filled ships, meaning that the American craft had to be built larger for the same performance. Since helium was expensive, they could not freely valve gas when light as could the hydrogen-filled Zeppelins. Not only did water ballast have to be conserved to prevent the release of gas, but water was recovered from the exhausts of the gasoline-burning engines to compensate for the weight of fuel consumed, the water recovery apparatus adding considerable weight. Much of the time the American ships were obliged to fly heavy or light at angles unacceptable in German practice.

One flight to South America in the *Graf* was very much like the other, and I do not propose to go into detail. The last of the six was the so-called "Christmas flight" (December 8–19, 1934), an event in itself as the *Graf* in the process completed 1,000,000 kilometers (621,000 miles) in the air, a record never achieved before or since by an airship. We stopped at Seville on our way back from Pernambuco to try out the new mooring mast. Practically all the top brass was aboard, and I have a postcard, written to myself on this

Dick mailed this postcard to himself on the 1,000,000-kilometer flight of the *Graf Zeppelin* in December 1934. The signatures are those of Max Pruss, who later became commander of the *Hindenburg;* Captain Lehmann, who also commanded the *Hindenburg* and headed the Reederei; Dr. Eckener; Captain Flemming, commander of the *Graf Zeppelin*; Captain Wittemann, also commander of the *Graf Zeppelin* and one of the survivors of the *Hindenburg* catastrophe; and Captain von Schiller, also commander of the *Graf Zeppelin.* (From the collection of Harold G. Dick)

particular flight, with the signatures of Dr. Eckener, Captain Lehmann, Captain Flemming, Captain Wittemann, Captain von Schiller, and Max Pruss, who at that time was still a navigator and not yet a captain.

Back in Friedrichshafen, progress on the new ship—the LZ 129, later named the *Hindenburg*—was slow during the winter of 1934, and in July 1935, George Lewis and I returned to the United States.

Chapter 9 *Mountain Climbing in the Alps*

Mountain climbing fascinated me more and more after I returned alone to Friedrichshafen in October 1935. Since this was a nice, neutral topic I frequently wrote home of my adventures, causing some concern to President Litchfield who cautioned me not to "go in *too strongly* for mountain climbing, even though you think it is safe," and kidded me about transferring me to Goodyear's Calcutta office "so you can climb Mount Everest." By the winter of 1936–37 my motto was "a mountain each weekend" and by February 5, 1937, I had made good my promise with a mountain climbed each weekend for six weekends running.

I had promptly joined the German-Austrian Mountain Club, which was made up primarily of Germans who enjoyed hiking or climbing in the mountains in the summer and skiing in the winter. Each area or chapter had a hut in the mountains. Some of the huts were very small, capable of accommodating perhaps only twelve men. Others were quite large and pretentious, accommodating many more. As an example, the Ulmerhütte, the hut of the City of Ulm chapter, could accommodate about one hundred people during the 1930s. Quite a flourishing enterprise, it was located about

The Ulmerhütte, altitude 7500 feet, was a flourishing enterprise. Located about 1600 feet above the top of the Arlberg Pass, it could accommodate 105 persons in the 1930s. (Photo by Harold G. Dick)

1500 feet above the top of the Arlberg Pass and Saint Christof. It was about 1500 feet above timberline.

The Friedrichshafen hut was rather small and unpretentious, capable of accommodating about twenty men. It was well above timberline, and had only a single caretaker. In the late fall all possible food and supplies were packed in to the hut and only bare necessities were packed in during the winter. Meals were the simplest imaginable, with only one item on the menu for any one meal. We normally carried our own tea, sausage, bread, and butter with us and depended on the hut to provide us with hot water so we could make our own tea. At that time a liter of hot water cost one groschen, just a few cents. We all drank tea to replenish the moisture in our systems for one becomes terribly dehydrated climbing in the mountains, particularly in winter. Beer, wine, and alcoholic drinks were all frowned upon as having no place when one was active in the mountains.

Lodging in these mountain huts was extremely cheap. If one belonged to the mountain club, a night's lodging cost 15 to 20 cents, or double that if one did not belong. In the smaller huts, which were the ones we frequented the most, the accommodations were far from luxurious. The one sleeping room would accommodate fifteen to twenty men. Each man got a straw-stuffed mattress and a blanket; they were all stacked up side by side. One blanket was sufficient for there usually was only one small window in the room and there was more than enough warmth. It was always a relief to get back out in the fresh air after spending the night in one of these places.

Primitive as some of these mountain huts were, however, they were adequate for what we were doing. A weekend in the mountains in wintertime was primarily an opportunity to climb a mountain on skis. We would leave Friedrichshafen by train or car as soon as my friends could get away from the Luftschiffbau or the Dornier works, on a Saturday afternoon. Arriving at the end of the road or where we would leave the train, we would start up into the mountains, arriving at the mountain hut anywhere from 10 p.m. to perhaps 2 a.m., after climbing from three to five hours on skis with skins attached to the skis much as the cross-country skiers do today. We would be on our way again in the morning before sun-up and would stop at the hut again only in the latter part of the afternoon on our way down into the valley. During the day it was very rare and quite exceptional for us to see anyone other than those of our own party, usually made up of four or five people.

The gain in altitude on Saturday evening varied from about 3000 to 5000 feet, the first 1500 feet usually below timberline, and the rest above timberline. It made for a strenuous weekend but by doing this each weekend one became conditioned to it. In the winter of 1936–37 I was in the mountains each weekend from the first of December to the middle of May without missing a single weekend and spent a week in these mountain huts over the Christmas holidays and again at Easter. The skiing was good through April, and even into May if one went up to about 6000 feet.

Dr. Eckener liked to kid me about this activity. He said I did not go skiing, I was out climbing a mountain on skis. He was right, but we *did* have some great long runs in virgin snow from the tops of the mountains back down into the valleys where we would catch the train back to Friedrichshafen.

Even in the mountains one could not escape the Gestapo. Several of us were out skiing and were stopping at one of the German-Austrian Mountain Club huts in the vicinity of the Austrian-Swiss border. This hut was at about 8000 feet, 2000 feet above timberline, timberline in the Alps being at about 6000 feet. The border between Austria and Switzerland in this particular area was about 10,000 feet or another 2000 feet above the Mountain Club hut.

In the hut we found a member of the Gestapo with an Austrian he had taken into custody. He was in the process of taking the Austrian back down to the little town below. The Austrian, from what we were told, had planned to go up into the mountains, cross over into Switzerland, and escape by going down the mountains on that side of the border. To do this he had hired a local man to take him up into the mountains and with luck get him to the border and over into Switzerland, using a sled and skis.

Word got about in the little Austrian village as to what was taking place and a report was made to the Gestapo. They sent an agent who was familiar with the mountains after the Austrian. The agent overtook him somewhere above the 8000-foot level and brought him back down as far as the Mountain Club hut where we encountered them.

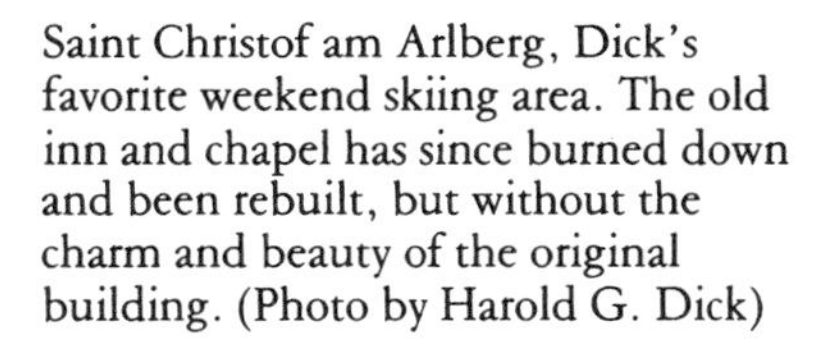

Saint Christof am Arlberg, Dick's favorite weekend skiing area. The old inn and chapel has since burned down and been rebuilt, but without the charm and beauty of the original building. (Photo by Harold G. Dick)

When we got to the hut in the late evening the Gestapo agent and the young Austrian were seated together drinking tea and were not going to continue down until the next morning. One can only wonder what ever happened to the young Austrian after that.

Several times when I returned to Friedrichshafen after skiing in Austria, wearing ski clothing with a rucksack and skis, the border police would ask many questions, such as where was I going, what was I doing in Friedrichshafen, how long would I be there, when was I going back to the States, and so on. Most of the time they appeared satisfied with a few simple answers, but on one occasion such was not the case.

This time the questioner was not one of the regular border police. His uniform was different and his manner was entirely different. He was by no means pleasant, and in fact was rather arrogant and obnoxious. He asked more questions than any of the others ever had, a very different routine from that to which I had become rather accustomed. At the border crossing the train would stop. Everyone would leave the train, pass through the border control, and reenter the train at the other end of the car. It was a relatively simple procedure providing the train waited until everyone was back aboard.

While so many questions were being asked of me that night the train engineer must have become impatient and tired of waiting, for he started the train and proceeded to pull out of the border control station. I had no desire to be left standing there so I answered the last question by shouting as I ran out of the control station and jumped aboard the moving train that I could not return to the States wearing only ski clothing.

I half expected someone to follow me—or perhaps more drastic action to be taken—but nothing happened. I think my interrogator was so surprised that anyone would do as I had done that by the time he recovered, I and the train were well on our way to Friedrichshafen.

It was during the winter of 1934–35 that my interest shifted to the *Hindenburg* and the problems of conserving helium, it being expected then that the nonflammable gas could be obtained from the United States.

Chapter 10 *The Design and Construction of the* Hindenburg

or all her pioneering achievements during an eight-year career, the *Graf Zeppelin* is not so generally remembered today as is her younger sister, the *Hindenburg*. This is because of the *Hindenburg*'s outstanding size, the incomparable spaciousness and luxury of her passenger accommodations (referred to often today in contrast to the "high density seating" on modern jet airliners), her spectacular achievement of being the first to offer regularly scheduled passenger service over the North Atlantic, and lastly, the fiery disaster at Lakehurst on May 6, 1937, in which thirty-five persons died, recorded in all its horror on film shown again and again through the years. The *Hindenburg* was very much a part of my life in Friedrichshafen: I was sent over to report on her design, I followed her construction almost from the beginning, participated in all but one of her early flights, and was aboard for two transatlantic journeys in 1936 to Rio and back, and for three journeys to Lakehurst. I will explain later why I was *not* on board for the first North American flight of 1937, which was also the last.

The *Hindenburg* represented Dr. Eckener's ideal intercontinental airship, and was intended to carry fifty passengers in luxury and comfort in a regularly scheduled service across the North Atlantic. While the *Graf Zeppelin* had been built as a demonstration passenger airship, she was too small and slow for regular North Atlantic service, this in fact a result of the dimensions of the 1916 hangar in Friedrichshafen which set upper limits to length and height, and therefore gas volume.

Larger ships were possible after the construction in 1929–30 of a new building shed in Friedrichshafen financed by loans from the national government plus 2.2 million marks ($530,000) from the State of Württemberg. The new hangar measured 820 feet long, 164 feet wide, and 151 feet high, "clear inner dimensions." Yet the corresponding design for a large passenger airship was not that of the *Hindenburg*. The unnamed LZ 128, I understood, would have been 761 feet long and 128 feet in diameter, would have had four power cars each containing two Maybach VL-2 engines back to back, and would have carried thirty to thirty-four passengers. With 5,307,000 cubic feet of hydrogen, she would have been able to fly the North Atlantic in regular service.

LZ 128 would thus have been about the same size as the two British giants, R 100 and R 101, completed in 1929 for passenger service to Canada, India, and Australia. But on her first long-distance flight to India, on October 5, 1930, R 101, inflated with hydrogen, crashed and burned at Beauvais in northern France. Forty-eight of the fifty-six persons on board died, including the British Secretary of State for Air, Lord Thomson of Cardington. This

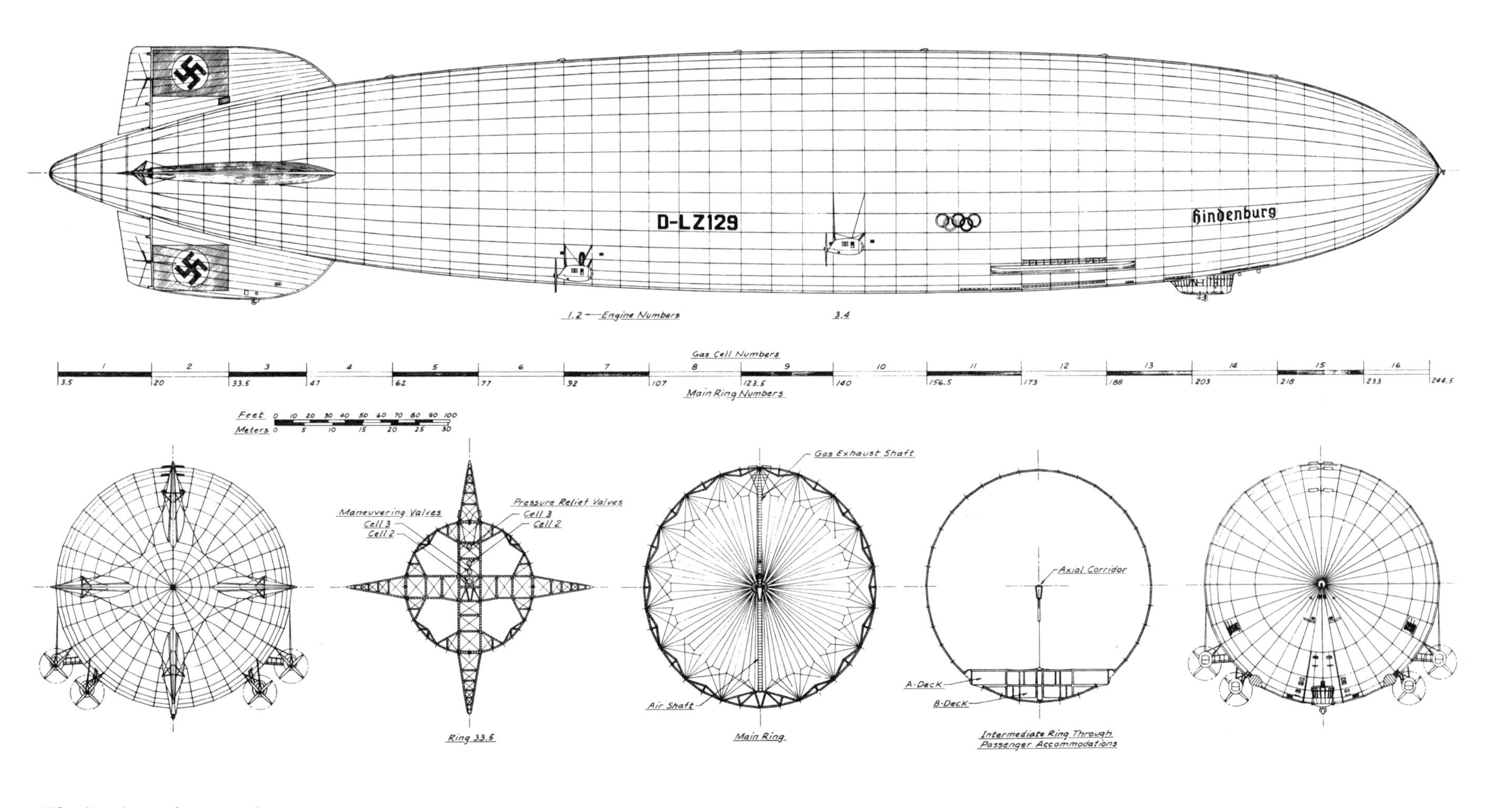

Side elevation and cross sections at designated rings or frames in the *Hindenburg*. (Drawings by William F. Kerka)

tragedy, and the increased production of nonflammable helium in the United States (though the Helium Control Act of 1927 forbade its export) persuaded Dr. Eckener to abandon the LZ 128 design. The LZ 129 would be larger, to obtain the same performance with helium.

The LZ 129 design called for an airship 803.8 feet long,[1] 135.1 feet in diameter, with a gas volume of 7,062,000 cubic feet. She was to have accommodations for fifty passengers, a crew of forty to forty-five, and a cruising speed of 77 MPH. Again, for the sake of safety, the inflammable gaseous fuel, Blau gas, would not be used. Instead, diesel engines would be fitted, burning a light oil (from the Dutch island of Aruba, in the Caribbean) which had a much higher flash point than gasoline.

In fact, the *Hindenburg* benefited from the rearmament program just commenced by Adolf Hitler: the Daimler-Benz firm of Untertürkheim near Stuttgart had developed a sixteen-cylinder high-speed diesel, the MB 502, with a maximum output of 1320 HP, for use in motor torpedo boats designed to make 36.5 knots. A lightened version, the LOF 6, with a dry weight of 4348 pounds, delivered a maximum of 1300 HP for five minutes at takeoff, and 850 HP for cruising. Compressed air was used to start the engine; it could be reversed by stopping the engine, shifting the cam shafts with compressed air, and then restarting the engine while rotating it in the opposite direction. Four of these power plants were fitted in the *Hindenburg*, and after a few problems in the early flights, gave excellent service. There were also two 50-HP diesel engines in a room in the keel at frame 145.5, driving two 220-volt generators for the ship's electrical system.

The hull was built up of fifteen main frames or rings, each consisting of eighteen diamond-shaped trusses arranged end to end in a circle, with reinforcements at the bottom in the region of the keel and adjacent to the gondolas at frames 92.0 and 140.0. Nine of the frames were spaced 15 meters (49.2 feet) apart, with the four largest-diameter frames amidships 16.5 meters (54.1 feet) apart. Wire bracing radiated from fittings in the center of each frame to the eighteen joints of the diamond trusses. Eighteen longitudinal girders connected the main rings at the junctions of the diamond trusses. Eighteen more longitudinal girders tied together the apices of the trusses. There were two light unbraced intermediate frames between each pair of main frames. These intermediate frames were designed to reduce the bending loads on the longitudinals, and to support the outer cover. All the rectangular spaces formed in the hull by the intersection of frames and longitudinals were braced diagonally with hard-drawn steel wire to take shear loads. The triangular keel in the bottom of the hull ran from the nose to frame 47 at the forward edge of the bottom fin. It served not only as a passage way for access to all parts of the ship, but also as a means for carrying concentrated loads, such as 660-gallon fuel oil and water ballast tanks,[2] spare parts, and cargo which sometimes included an airplane or an automobile.

Lateral gangways led up frames 92.0 and 140.0 to give access to the power

The *Hindenburg*, LZ 129, over the airship hangars at Friedrichshafen.

Work on the outer cover and the huge fins of the *Hindenburg*.

The load-carrying keel of the *Hindenburg,* showing the maze of wires and controls, and the 660-gallon fuel tanks.

One of the power cars of the *Hindenburg.* The four-bladed wooden propeller is 20 feet in diameter. There was an engine mechanic stationed in the power car at all times. The engine was a 16-cylinder diesel geared down 2:1 and reversible to obtain reverse thrust during the landing maneuver.

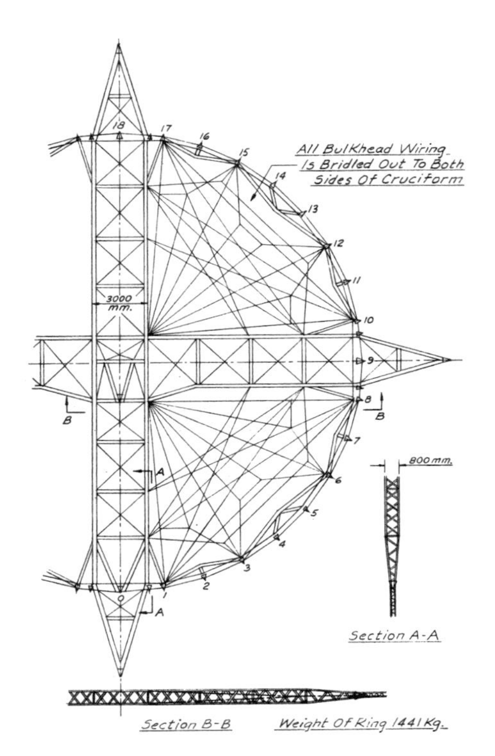

Main frame 47 in the tail of the *Hindenburg*. Shown here is the ring structure of duralumin girders and bracing wires, together with the cruciform structure going right through the ship as an integral part of the fins. (Drawing by William F. Kerka)

cars hung on these frames, and at the power cars were "stub keels" one bay long, holding service tanks for diesel fuel, lubricating oil, and spare radiator water. The streamlined gondolas, each 20 feet long, were attached to the hull by struts and suspension wires. They were staggered so that the propeller slipstream of the forward cars, which were higher on the ship, did not strike the after ones. The forward cars were mounted at 4 degrees to the ship's axis; the after ones at 3 degrees, to direct the propeller blast outward and away from the hull and to minimize damage to the outer cover. Each car housed a mechanic, with room for more personnel to work on the engine in case of a breakdown. At the rear of the gondola was a large four-bladed fixed-pitch wooden propeller (actually two two-bladed propellers fastened together) 19.7 feet in diameter, and driven through reduction gears with a 2:1 ratio.

At the stern were the four large fins, 105 feet long, 49 feet in breadth, and 12 feet thick at the root, tied into the ship's structure by heavy cruciform girders at main frames 20.0, 33.5, and 47.0. There were auxiliary rudder and elevator controls in the bottom fin, and since there was no centerline engine car aft, the stern of the ship was supported on the ground by a swiveling air wheel in the bottom of the fin which could be partially retracted in flight.

The control car, forward at frame 203.0, was much smaller than that of the *Graf Zeppelin*, as the *Hindenburg*'s passenger quarters were inside the hull. The control car included three compartments: the control room forward, a chart room amidships, and a room aft for the navigator to take bearings for the dead reckoning type of navigation described earlier.[3] Under the control car was another swiveling air wheel identical to that under the tail. The radio room was in the keel immediately above the control car, and the captain's cabin a short distance forward. The control room was the duty station of the officer of the watch, his navigator, and the rudder and elevator men. The latter had more sophisticated equipment than in the *Graf*. There were ⅓ HP servo motors for both the elevators and rudders, used in smooth air. The automatic rudder control (slaved to the gyro compass) was intended to keep the *Hindenburg* on course with less deviation than the human helmsman. (This was not satisfactory in 1936, but was effective in 1937.) If turbulence were encountered, however, the automatic system was disconnected and the elevators and rudders controlled by hand.

The nonbarometric altimeter, instead of using a shotgun cartridge as in the *Graf Zeppelin*, featured a whistle operated by compressed air, the whistle being near the bow at station 228 and the receiver abaft the control car at station 188. At the elevator man's stand there was a gas cell pressure board with indicators for each of the fourteen gas cells. A red light showed when the cell was at pressure height and 100 percent full, while individual pressure indicators read from 0 to 20 mm of water. A ballast board had red and green indicators for the four emergency bags aft at frame 47 and four at the bow at frame 218, each emergency bag containing 1100 pounds of water. Also, the watch officer and the elevator man could see at a glance how much water was

The control car of the *Hindenburg*, which was much smaller than that of the *Graf Zeppelin*. The forward section was the control area or bridge. The center section was the navigation or chart room, and aft of this was a small utility area.

Elevator (*left*) and rudder controls, shown here actually of the LZ 130 where they were the same as in the *Hindenburg* with the exception of the automatic control at the rudder station. The ballast control toggles are at *upper left*.

There was an emergency control stand in the *Hindenburg* in the lower fin, with elevator and rudder wheels and a minimum of instrumentation.

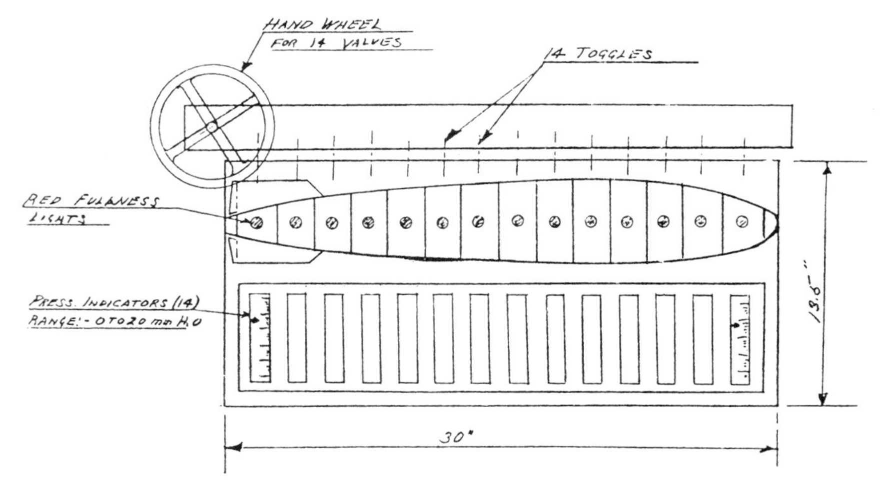

The gas cell pressure and maneuvering valve board in the *Hindenburg*. (Drawing by Harold G. Dick)

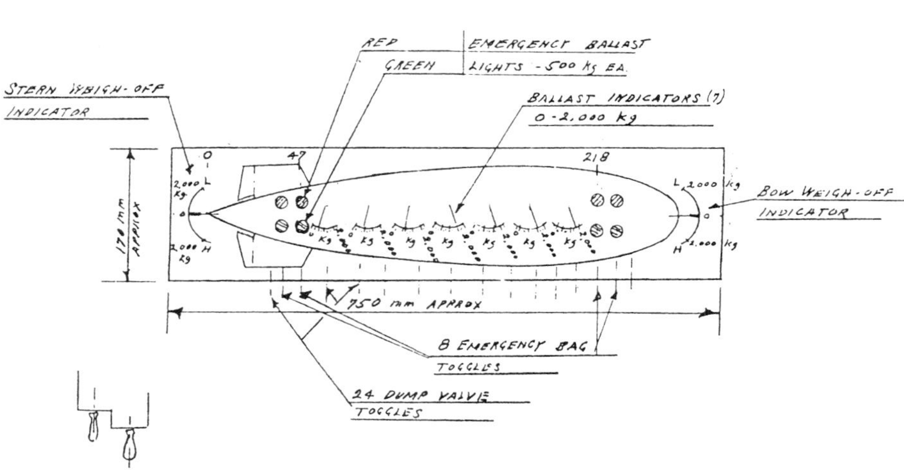

The ballast board at the elevator man's stand in the *Hindenburg*. (Drawing by Harold G. Dick)

present in each of seven ballast tanks, of 4400-pound capacity, spaced along the keel. The ballast board also had bow and stern weigh-off indicators—reading from 4400 pounds light to 4400 pounds heavy—which were very useful in weighing off the ship before takeoff.

The *Hindenburg* had a gyro compass system, the "mother compass" being in the control room of the generator station in the keel, and the "daughter compasses" (as the Germans say) being at the rudder stand in the control car, on the port and starboard sides of the navigation room, with the radio compass equipment in the navigation room, and in the after part of the control room. In the control room there was a telephone exchange connecting with fourteen locations in the ship, with a separate line to the auxiliary station in the lower fin. Orders could be given to the sailmakers in the axial gangway via a speaking tube, thus eliminating electrical equipment which might have posed a fire hazard. On the starboard side of the control room were the engine telegraphs, while one of the telephone connections was with the engineers' room in the keel which was the duty station of the engineer in charge of each watch of machinists.

While there were many sophisticated features of the *Hindenburg*, as befitted the 118th ship built by the Luftschiffbau Zeppelin, the basic design and structure were thoroughly conventional, presenting features introduced as far back as 1915 and 1916. All the structural girders in the ship (except in the passenger area) were of conventional triangular cross section, built up of drawn duralumin channels and stamped spacers, with girders of different dimensions and thickness of metal used in places where more or less stress and loading would occur. The main frames were of conventional flat "bicycle wheel" layout, with radial bracing from a central fitting. The thirty-six longitudinals ran from the nose cap to frame 20, where they were brought together; from frame 20, the rudder post, into the tail cone, there were only twenty in number. The cruciform bracing of the fins went back to 1915 when the cruciform structure was introduced at the rudder post. The thick cantilever fins dated from 1918, and the second cruciform at the thickest cross-sectional area was introduced in the *Los Angeles* of 1924. Moreover, Dr. Eckener and Dr. Dürr were unwilling to dispense entirely with fin bracing: stout cables with a breaking strength of 40,000 pounds connected all four of the fins at the rudder post with the hull at frame 20. I emphasize these conventional and conservative features as they contrast with the many innovations in Goodyear-Zeppelin's two big airships (discussed in Chapter 18), the *Akron* and *Macon* of 6,500,000 cubic feet.

The navigation or chart room in the *Hindenburg:* there were two gyro compass repeaters, one on each side, a radio compass, and the telephone switchboard.

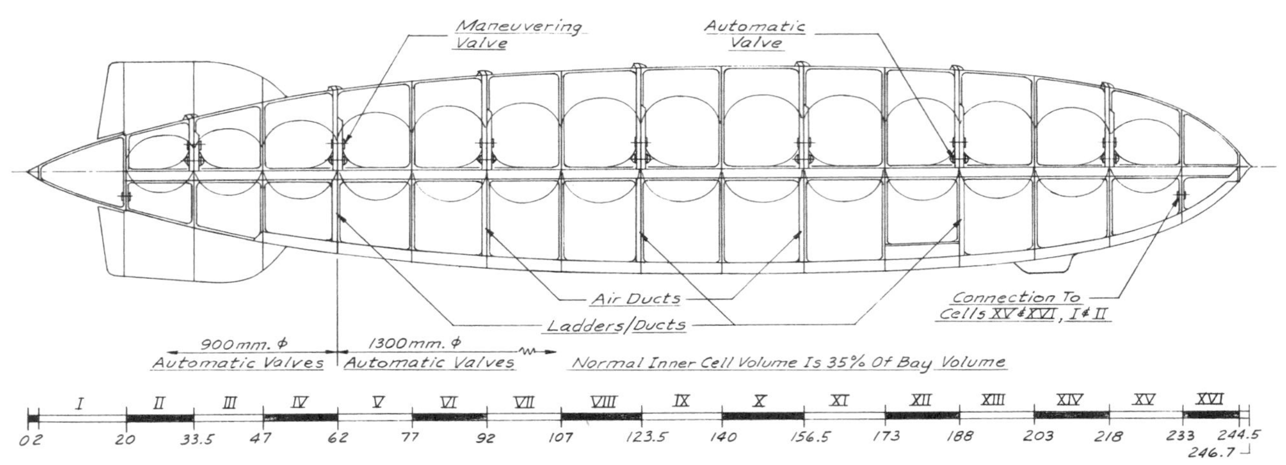

Arrangement Of Gas Cells LZ-129

ON PAGE OPPOSITE, TOP:
Proposed arrangement of gas cells in the *Hindenburg* with small hydrogen cells inside larger helium cells. (Drawing by William F. Kerka)

BOTTOM:
The axial corridor that extended from bow to stern on the central axis of the *Hindenburg:* this was a load-carrying member supporting the bulkheads in the event of a deflated cell. The axial corridor also gave access to the gas cell valves.

The gas cell arrangements of the *Hindenburg* were novel, with features deriving from an early scheme to inflate the sixteen main gas cells with helium, while inside fourteen of the helium cells would be a smaller cell, about 35 percent of the volume of the helium cell, which would contain hydrogen. The hydrogen could then be valved off to bring the ship into equilibrium as fuel was burned, while the expensive helium could be retained. Since the small hydrogen cells were high up in the ship, their valves (these were to have had only hand-operated maneuvering valves, no automatic ones) could not be monitored from the keel in the bottom of the ship. This required the fitting of an axial corridor whereby the sailmakers could have access to the valves, while the maneuvering and automatic valves for the big helium cells would also be installed at the axial corridor level. A conventional ventilation shaft led upwards from the valves between every other pair of gas cells to the top of the ship, where streamlined hoods open to the rear created a vacuum to suck gas out of the ship.

Late in the construction of the LZ 129 it was realized that helium would not be available and the inner cells were never installed, but by then the Luftschiffbau Zeppelin was committed to the axial corridor with the valves in the centerline of the ship.

The axial corridor ran from one end of the ship to the other, was triangular in cross section, and was high and wide enough for a man to walk upright. It was an important stress member, relieving loads on the radial wiring of the main frames and reducing deflection of the bulkhead in case of a deflated gas cell. The corridor was permanently fixed in the nose and tail sections, and the corresponding gas cells, numbers 1, 2, and 3 aft, and 16 forward, were split so they could hang down on either side of the corridor. Between frames 47 and 223, however, the axial corridor was removable, with a 15 or 16.5 meter section corresponding to the length of the bay, and going through a tunnel in the gas cell designed to accommodate the corridor segment. Ladders from the lower keel gave access to the axial corridor at frames 62.0, 123.5, and 188. In the completed ship, both the maneuvering and automatic valves were housed in a cage just above the axial corridor, and were accessible from it. In all, there were fourteen automatic valves (the foremost and aftermost cells had sleeves connecting them with the adjacent cells). Also there were fourteen maneuvering valves which could be opened simultaneously with a wheel in the control car.

The gas cell material in the *Hindenburg* was novel. Years of research designed to develop a substance as impervious as expensive goldbeater's skin had paid off, and in the *Hindenburg* a gelatine solution resembling the gelatine latex developed for gas cells in the United States was brushed on in six coats to render the cotton fabric gastight. The weight of the fabric was 180 to 200 gm per square meter. As in earlier airships, the lift of the gas cells was transferred to the structure via large-mesh gas cell wiring, and small-mesh ramie cord netting. The latter was waterproofed where it ran between the top backbone girder (number 18) and the main longitudinals adjacent

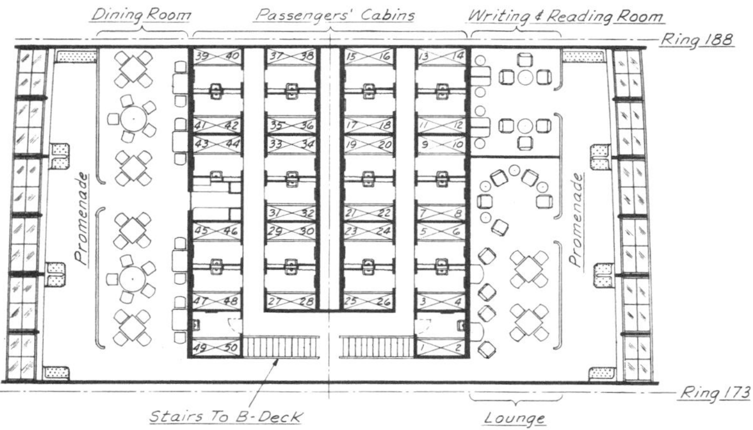

Plan of A Deck in the *Hindenburg*.
(Drawing by William F. Kerka)

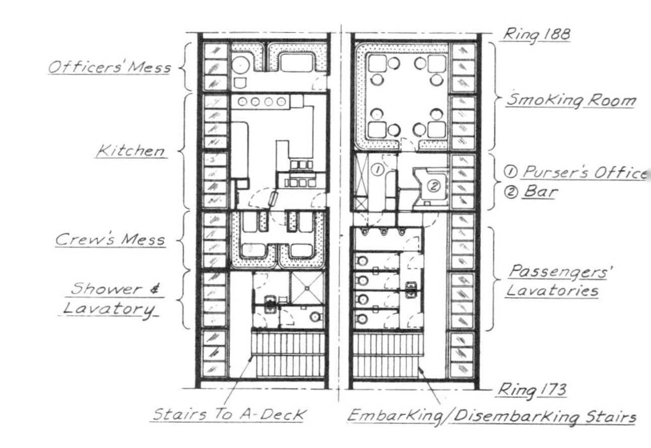

Plan of B Deck in the *Hindenburg*.
(Drawing by William F. Kerka)

ON PAGE OPPOSITE, TOP: The scaffolding for the passenger quarters of the *Hindenburg* during construction. The passenger quarters were entirely within the hull and on two levels.

(number 17 right and left), as it had been found in the *Graf Zeppelin* that moisture conducted by the ramie cord corroded the duralumin girders.

The *Hindenburg* was equipped for either mechanical or manual ground handling: Luftschiffbau Zeppelin personnel, in contrast to the U.S. Navy's, preferred to fly the ship to the ground and walk it to the mast, rather than to make a "flying moor" with the ship approaching the mast in the air and then being hauled down by the main mooring line. A new telescopic traveling mast had been developed to take both the *Hindenburg* and the *Graf Zeppelin*, with a maximum height of 70.36 feet for the former and 55.12 feet for the latter. The mast traveled on two tracks twenty feet apart and was braced to side trolleys by cables. The winch for the main mooring wire was hand operated—something that would not have been tolerated in the United States! While at the mast at Lakehurst, the *Hindenburg* was positioned with the lower fin at station 42.5 attached to a riding out car which ran on a circle of standard-gauge railroad track, so that the ship could vane into the wind. There was some concern that a gust of wind might cause the *Hindenburg* to lift the car right off the tracks, so the car, weighing seven tons, was loaded with an additional three tons of lead bars. Ground handling with no mechanical aids was much as with the *Graf Zeppelin*, though with a larger number of men. The ground crew for landing at Lakehurst in 1936 totaled 247 men.

The passenger quarters in the *Hindenburg* are the feature of the ship that people ask the most questions about, and as the years pass, they seem ever more magnificent and spacious. I even read that every one of the two-berth cabins had its own private bath and toilet![4] This certainly was not true, but even so, the *Hindenburg*'s quarters far outshone the cramped accommodations aboard today's long distance jet aircraft.

The *Hindenburg*'s passenger quarters were on two levels and for the sake of both space and streamlining were entirely inside the ship's hull. They were in the bottom of bay 12, but occupied only a fraction of the space in the bay, cell number 12 still having 80 percent of the volume of the adjacent cell 11. The upper level or A deck was the larger, measuring 46.6 by 92 feet, and it included all of the sleeping cabins and public rooms. The lower level or B deck included toilets, a shower, the galley, and a small bar with an airlock door that permitted entry into a smoking room, an unusual feature of a hydrogen-inflated airship. Entrance to the passenger quarters was via a pair of stairlike gangways that could be lowered to the ground with a winch; at the mast at Lakehurst, these gangways were fitted with a set of bicycle wheels so they could roll around the mast with the ship.

A further flight of stairs both port and starboard took the passengers to A deck, where they found a bronze bust of Field Marshal Paul von Hindenburg, for whom the airship was named, in a niche on the forward bulkhead. In the center of A deck, and leading onto two corridors, were the twenty-five sleeping cabins. These measured only 78 by 66 inches and had no outside exposure. Each had an upper and lower berth, a tip-up wash basin with hot

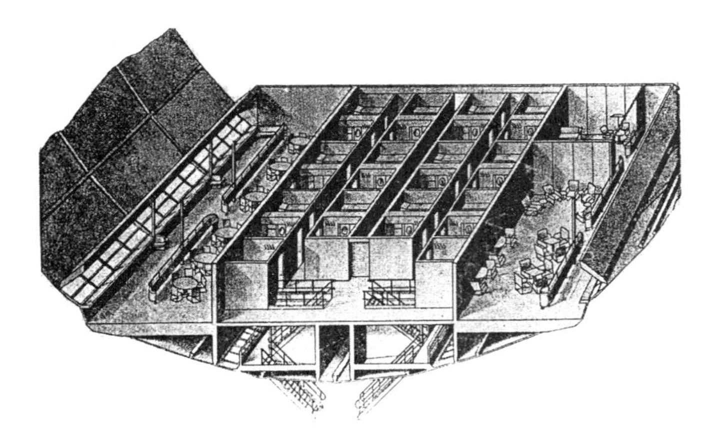

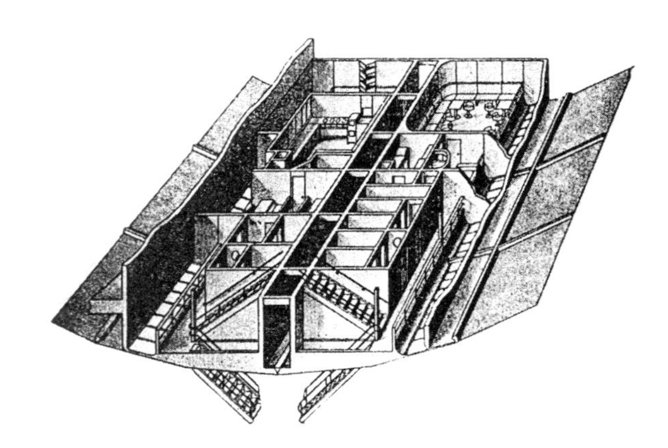

Passenger accommodations in the *Hindenburg:* A Deck (*left*). B Deck (*right*). (From the collection of Harold G. Dick)

and cold running water, a collapsible writing table, and a call signal for the steward. The cabin walls were lightly built of two thicknesses of airship fabric with a layer of lightweight foam in between, and were colored light blue, gray, or beige. The sliding cabin doors, of the same material, weighed only 6.8 pounds each with the latch.

It was not expected that the passengers would spend their waking hours in the cabins,[5] so much attention had been given by Dr. Dürr and particularly by the interior architect, Professor Fritz Breuhaus, to the public rooms on A deck. On the port side, outboard of the sleeping cabins, was the dining room, occupying a space 46.6 by 12.7 feet. Here all fifty passengers could eat at one sitting, at tables covered with fine linen and decorated with fresh flowers, dining with real silverware from dishes especially created for the *Hindenburg*. The design of the china showed the Deutsche Zeppelin Reederei crest, a white Zeppelin outlined in gold, superimposed on a blue globe with the meridians of longitude and parallels of latitude in gold. With the head chef a former staff member of the Kurgarten Hotel, the meals were unsurpassed, while the *Hindenburg*'s wine cellar (actually a storeroom in the keel) would have done credit to a nobleman's castle.

The corresponding space on the starboard side was unevenly divided into a lounge and a writing room. The lounge, at the insistence of American passengers, boasted a grand piano (covered in yellow pigskin) whose weight, through the use of duralumin, was kept down to 397 pounds.[6] Outboard of the public rooms, and separated from them by a low rail, were promenade decks 46.6 feet long and 5.7 feet wide, connected by a cross-passage at the after end of A deck. Outboard of the promenade decks were six large Plexiglas windows, slanting outward at 45 degrees. Three of these windows could be opened on each side as there was no draft in or outwards with the ship under way. People aboard spent hours at the windows gazing down on the landscape or seascape passing close underneath as the ship usually cruised at an altitude of 800 feet or so. The public rooms were decorated with mural paintings by Professor Arpke; the twenty-one in the dining room, for instance, depicted scenes on the *Graf Zeppelin*'s route to South America.

Unfortunately no color reproductions of these original art works were made, or if they exist, I am not aware of it.

In contrast to those of the *Graf Zeppelin*, the passenger quarters of the *Hindenburg* were heated, the source being the forward main engines. Some of the hot water from the engine cooling systems was led to a set of radiators near the passenger quarters, where electric blowers forced air through the radiators and into ducts leading to the cabins and public rooms. With a water temperature of 176 degrees F, the air temperature was 169 degrees F, which could be lowered with cool air as necessary.

On the lower level or B deck were the more prosaic features of the passenger quarters. On the port side was the electric galley, including an electric stove with four burners, electric roasting and baking ovens, a refrigerator, and an ice machine. A dumbwaiter carried prepared meals up to the dining room. Forward of the galley was the officers' mess, and abaft it, the crew's mess, both off limits to passengers. Then came a shower, with an adjacent toilet. Passengers signed up to use the shower, the first such luxury aboard any aerial conveyance. Designed to conserve water, it turned itself off after running for a certain predetermined length of time, perhaps leaving the passenger all soaped up. Waste water from the shower, along with that from the washbasins, was collected and saved as ballast in dirty-water tanks as in the *Graf Zeppelin*.

On the starboard side of B deck, from aft to forward, were the ladies' toilet with two stalls, the men's toilet with two stalls and three urinals, and the chief steward's room with bunk, slightly larger than the passenger cabins. (The only way, incidentally, that the passengers could reach the keel, which ran through the center of B deck, was through the steward's room, and it was he who conducted small parties through the ship on request.) Next to his room was the tiny bar measuring 6 by 7.8 feet. The only entrance to the smoking room, which was pressurized to prevent the admission of any leaking hydrogen, was via the bar, which had a swiveling air-lock door, and all departing passengers were scrutinized by the bar steward to make sure they were not carrying out a lighted cigarette or pipe. The smoking room measured 12.6 by 15 feet, and to minimize the fire risk, had yellow pigskin leather on the walls, as well as on the settees which ran around three sides of the room. The fourth side had a low rail beyond which the occupants could look straight down to the ground passing underneath—for B deck, and the crew's quarters fore and aft, were illuminated by a long row of Plexiglas windows 48 inches wide, set flush in the bottom of the ship.

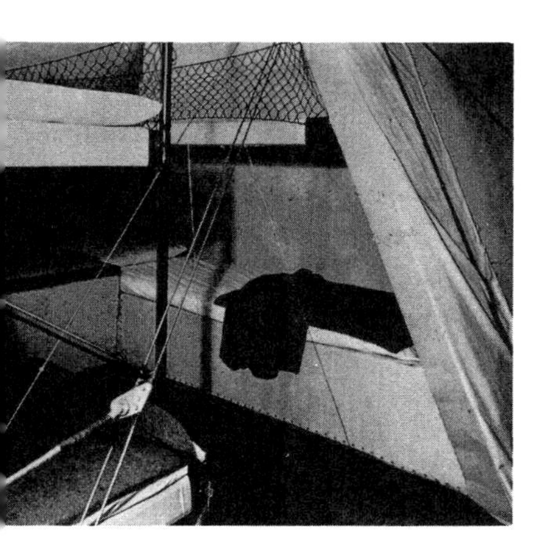

Crew quarters in the *Hindenburg*.

The *Hindenburg* crew was accommodated in three locations in the keel: Officers' quarters, including bunks for twelve officers and the captain's private stateroom, were in bay 14 just forward of the control car. The deck force had twenty-two bunks in bay 11 just aft of the B deck passenger quarters. The machinists had twelve bunks in bay 5 well aft in the keel.

In May 1934, when I arrived in Friedrichshafen, the construction status of the *Hindenburg* was as follows:

The hull structure was complete as far back as the leading edge of the fins. The foremost cruciform frame, number 47, was under construction. About ten meters of the stern assembly (that portion aft of frame 20, the last cruciform frame) was complete.

Only one fuel tank had been mounted in the ship. Passenger acommodations were about 50 percent structurally complete. Flooring and plumbing for hot and cold running water was installed.

The Daimler-Benz engines were still on the test stand in Stuttgart, having been run about one hundred hours. A water recovery set-up using silica gel was complete, with a unit being one-third the size of that required for a 1000 HP engine.

Crew's and officers' quarters were structurally almost complete, but without walls, fittings, plumbing.

None of the outer cover was in place. No gas cells were installed, but a test of a hydrogen inner cell was scheduled for the next few days. Construction of the control car was under way, but with no installations. One diesel engine for the electric power system had been tested, but there was no installation in the ship. Nothing had been done with the control system or mooring equipment. The ship was a long way from completion, and it was still being assumed that the ship would be inflated with helium.

Whereas German airship operators in the past had been willing to valve off cheap hydrogen to compensate for the burning of gasoline on long flights, this would not be possible with the *Hindenburg* inflated with helium. Some means would have to be found to keep the all-up weight of the airship more or less constant while fuel was consumed. The Germans rejected the American solution of recovering water from the exhaust of the engines. Their reasoning was that the recovery apparatus performed inefficiently, being mounted in the boundary layer on the surface of the hull, while it was heavy and required constant cleaning to remove soot accumulated in the condensers. I find that even among airship historians little is known about the technology considered for the *Hindenburg*, presumably because the ship finally flew with hydrogen. Having access to all the secrets of the Zeppelin Company, I can list all the measures that were considered:

1. Recovery of water ballast by the use of silica gel to absorb moisture from the air.

2. Use of a hydrogen inner cell suspended inside the helium cell. The hydrogen "antiballast" would be valved to compensate for the weight of fuel consumed.

3. Use of a water pickup system that would pump sea water up into the ship.

4. Use of a supplemental system of rain gutters attached to the hull to catch water runoff in a rain storm.

5. Addition of a fifth engine being carried specifically to burn hydrogen, thus reducing the lift while developing some power and producing a pure form of water ballast.

Many of these developments were pursued during my first visit to Friedrichshafen between May 1934 and July 1935. During this period every possible effort was made to devise a suitable water recovery system using silica gel. Some success was had but with considerable complication. The gel was dropped into a vertical tower 16 feet tall, where it fell through a rising column of outside air, absorbing moisture. At the bottom of the tower the moist gel was extracted, exposed to the heat of the exhaust of a World War I Maybach MbIVa engine which vaporized the absorbed moisture, and this was then condensed while the dry gel was returned to the top of the tower. The apparatus took up space inside the ship, the weight of it was considerable, and it worked only in air with high water content.

By August 1934, using silica gel to absorb moisture from the air, a test setup with air at 68 degrees F, relative humidity of 67 percent, and air flow of 141 cubic feet per second, 8.8 pounds per minute of gel in the system produced 1.3 gallons of water in 23 minutes or 3.43 gallons per hour. Although a system that would extract 50 percent of the moisture in the air seemed practical, it was still insufficient since an absorption factor of 70 percent was desired.

In May 1935, final design for one unit to be installed in the ship was authorized. At this time it was generally agreed that LZ 129, inflated with helium, would operate with a silica gel recovery unit, a water pickup system, and a system of rain gutters on the ship. However, in November 1935, the silica gel system was referred to the Silica Gel Corporation, who offered to build a unit that would yield 330 pounds per hour for a weight of 3087 pounds. Additional equipment required in the ship would bring the total weight to 3969 pounds, while between 5 and 10 HP would be required to operate the unit. There was some question as to the Silica Gel Corporation's ability to meet such a guarantee, and unless the guarantee were met the unit would not be accepted.

The use of a hydrogen-inflated inner cell received a great deal of attention in 1934 with the first test installation being made in August of that year. Bay 10, with a total volume of 776,800 cubic feet, was chosen. The double test cell was made up from some of the old cells of the *Graf Zeppelin*, even though the material was practically six years old. The material, already old and weak, was allowed to lie in the test bay for some weeks beforehand so that the fabric was permitted to dry out even more. In a first testing procedure, the outer cell was ripped over a length of 50 feet during an attempt to position it during inflation. Following repairs, a second test was made a few days later. With 10,500 to 14,000 cubic feet of hydrogen in the outer cell and none in the inner cell, the assembly was floated into place in the bay with the help of the inflation net, and the central corridor segment, inserted in the tunnel of the outer cell, was pinned in place on the bulkhead fittings fore and aft. Some 2800 to 3530 cubic feet of hydrogen was then run into the inner cell to take the weight of the fabric off the corridor. This was all the gas that the inner cell could take, for the bag began to bulge out below the level of the gas in the outer cell; apparently the gas in the inner cell could not force its way upwards through the folds of fabric of the outer cell. The inflation of the outer cell was increased, and as positioning allowed, the inflation of the inner cell was also increased. When the inner cell had about 106,000 cubic feet, and the outer cell about 247,200 cubic feet, approximately 50 percent of the volume of the bay, the bulkhead attachment points were secured and inflation continued to 90 percent of the volume of the inner cell. Inflation of the outer cell was continued until the total gas volume was 70 percent of the volume of the bay. At this time it could be seen that the inner cell was completely free of the outer cell and very well positioned.

The inner cell was then deflated, but as both cells contained hydrogen, it was necessary to increase the inflation of the outer cell in order to force all the gas from the inner cell. (In the final arrangement, there would have been a pressure differential because of the greater density of helium.) This part of the test did not seem particularly good: as gas was valved from the inner cell, the inner cell settled down keeping the gas at the bottom, for with respect to the outer cell, the inner cell had no lift.

All of the observers considered the test satisfactory, but by the spring of 1935 the decision was that there would be no hydrogen-inflated inner cell on the *Hindenburg*'s first flight to the United States. This decision was probably based on the apparent success of the silica gel system for recovering ballast water. The water pickup from the ocean was considered quite feasible and tests were made with the *Graf Zeppelin* over Lake Constance. A more refined unit was to have been built in Hamburg but the water pickup system was abandoned when it was realized that LZ 129 would have to be inflated with hydrogen.

Rain gutters were mounted on the *Graf Zeppelin* in the summer of 1934 with a certain degree of success. They were likewise installed on the LZ 129 but were considered more of a convenience item to increase the fresh water supply in flight. Some flights produced absolutely no rain water so this system could only be considered as a supplemental system for obtaining water. Later, the *Hindenburg* succeeded in collecting as much as 5.5 tons of water in a few minutes from a heavy rain shower. It was common practice to "brush" the edge of a rain shower with the side of the airship in order to collect as much ballast as possible and reduce the amount of hydrogen that had to be valved.

From these various tests it can be concluded that the Germans seriously thought that the LZ 129 would be inflated with helium, but had to change their thinking during the latter part of 1935 or early 1936 when it became apparent that helium would not be available for the first inflation of the ship. Some thought was given to flying the ship to Lakehurst inflated with hydrogen, and then reinflating it with helium if this gas were made available at that time. There was actually some question as to whether some of the German operators really wanted to use helium because of their fine safety record with the *Graf Zeppelin* and other hydrogen-inflated airships, and because of the penalty of a 10 percent loss of lift from using helium. As construction of the LZ 129 continued, and it became clear that her estimated fixed weights would be exceeded, the possibility of using hydrogen appeared increasingly attractive.[7]

Actually it is difficult to understand just how Dr. Eckener expected to obtain helium for the LZ 129. Perhaps he believed that by publicly emphasizing his intention of using "safe" helium in the new ship, he would have some leverage on getting helium released by the Americans, though the Helium Control Act would have to be amended. How Dr. Eckener would have paid for the expensive gas I do not know, and the Nazis' tight control over foreign exchange would have frustrated him in any case. All this changed after the *Hindenburg* fire when it became a matter of prestige for the Nazi government to have the passenger Zeppelins flying with helium, but in the years 1934–36 there was no activity concerning the transportation, storage, and purification of helium such as I saw later in 1937–38.

Because the final decision had been made by Dr. Eckener to abandon the LZ 128 design and to substitute the larger LZ 129 to have the same performance with helium, I believe he was very disappointed in not having helium for the ship. From my close contact with Dr. Eckener during the nearly five years I spent in Friedrichshafen, I know that he was extremely safety minded, always weighing each decision against the worst possible results. To him, an *assumption* of correctness was inadequate; one had to *know* that one was right and only then proceed. With his vision, Dr. Eckener certainly realized that if the big rigid airship were to occupy the place in international commerce and transportation that he intended, it would have to operate with helium as its lifting gas.

On many of the 1936 flights of the *Hindenburg* we had representatives of the U.S. Navy aboard, and strangely enough, some of them adopted the thinking of some of the German operators that hydrogen was a safe gas because the Germans had learned the secret of how to handle it and accordingly there was no hazard involved in its use. On such occasions I would suggest to our naval observer holding these views that he and I should take a tour through the ship. We would climb one of the access ladders to the axial corridor which passed through a tunnel right in the center of the gas cells. When about halfway to the next access ladder, I would point out to him that there was about 65 feet of hydrogen on all sides of us and that if anything went wrong there would be no chance of getting out. The demonstration worked. None of those Navy men ever mentioned to me again the idea that hydrogen properly used was safe and should be used in any new or subsequent airships!

Chapter 11 *The* Hindenburg *Completion and Trials*

After three months in the United States I returned to Friedrichshafen in October 1935. Because of the chronic shortage of accommodations, I had to put up for a while at the Kurgarten Hotel. In contrast to the 80 marks per month that George Lewis and I had paid earlier for our rooms, here I was paying 8.50 marks per day, though this included breakfast. Early in November, however, the old rooms became vacant. This was fortunate, as another Goodyear employee and good friend, Karl Fickes, who headed the blimp operation, was supposed to come over for the first flight of the LZ 129. On November 19 I wrote to Karl that Dr. Dürr, Dr. Eckener, and his son Knut Eckener were agreed that inflation of the new ship would start right after New Year's so "if you plan to take a slow boat, and leave directly after the New Year you should be here in time to see the last steps of inflation."

I was also writing at this time to the U.S. Navy Lieutenant Commander Scott E. (Scotty) Peck, an old friend, who was slated to come to Friedrichshafen and fly in the LZ 129 as an observer. Since the ship was scheduled to fly regularly to Lakehurst, he was my source of information for the Luftschiffbau Zeppelin on the facilities and procedures there.

According to a letter that I wrote to Dr. Arnstein on December 9, 1935, the name *Hindenburg* had already been definitely chosen for the LZ 129 though she did not bear it during her early flights. Although it was now certain that the new ship would make at least her first flights with hydrogen, there was still the desire to recover water to compensate for the weight of fuel burned, thereby avoiding the need to valve hydrogen. There was still interest in silica gel, but a hydrogen-burning engine was now the center of effort. A wartime Maybach 250 hp engine was used in the tests and it was found that 35 to 53 cubic feet of hydrogen, burned in the engine, would yield 2.2 pounds of water. The engine would only develop one-fifth of its rated horsepower with gasoline, but the chief consideration was converting excess hydrogen to water ballast. Herr Sturm imagined that the ultimate hydrogen motor would be one of the big Daimlers generating a mere 300 hp, but producing per hour about 430 pounds of water "almost good enough to drink," compared to the 1350 pounds per hour of diesel fuel consumed. With the main engines throttled back and 10,500 cubic feet of hydrogen being burned per hour, the airship would acquire lightness at the rate of about 110 pounds per hour. A fifth gondola for the hydrogen motor for the LZ 129 was actually designed, but with the rush to get the ship flying, the hydrogen motor project was shelved.

Another technical innovation bandied about at the same time was a fixed trapeze or "perch" for hooking on aircraft. Captain Lehmann was pushing this scheme, with the ultimate aim of speeding the carriage of transatlantic mail: late letters would be flown out to the *Hindenburg* after her departure, while a hook-on aircraft would fly out from the destination, pick up the mail, and have it on the ground several hours before the airship's arrival. Lehmann wanted to have the perch on the airship when it arrived for the first time at Lakehurst, and during the three days she would be there, he wished to make a demonstration flight with a series of hook-ons and releases. On December 11, 1935, I wrote to Scotty Peck, on behalf of the Luftschiffbau Zeppelin, asking him to bring the drawings for the low perch fitted to the *Macon*. On February 28, 1936, I wrote to Dr. Arnstein pointing out that "no arrangements have as yet been made to obtain a plane with a hook or a pilot to fly the plane" in the demonstration flights, and suggesting that Goodyear handle the American end of the operation and get some possible publicity. The U.S. Navy in fact did release the perch drawings but it was not until the spring of 1937 that hook-on trials were made at Frankfurt with the *Hindenburg*, with results that I shall describe later (*see* Chapter 14).

Mr. Litchfield, Dr. Arnstein, Karl Fickes, Scotty Peck, and all my other correspondents in the States naturally wanted to know when the *Hindenburg* would be inflated, and when she would make her first flight. On November 22, 1935, I advised Dr. Arnstein that the ship still had considerable work to be done on her:

> The upper fin is complete with the exception of the covering, and the port horizontal fin has reached approximately the same stage. The starboard fin is not so far along for work is still being done on the hinges and on the structure at frame 20. A great deal of work is still required on the lower fin and I would estimate that it is less than 50% complete. The upper rudder is complete and has been fitted in place but will be removed to be covered. The port elevator is being fitted while the other two surfaces will not be ready for fitting until Christmas. The stern section, now two meters shorter,[1] has been erected. The outer cover has been carried as far aft as just forward of the fins although the belly of the ship has been left open. Nowhere has it received its fifth coat of dope. The two forward power cars have already been attached to the ship and the engine has been run in one of them. . . . The work of attaching the two other power cars is now in progress. Three engines are here and two more are promised in a week or two. . . . The new ship will not be inflated in December as it is considered undesirable to have the gas in the ship over the holiday when nothing much can be done. . . . Present plans are to start inflating the ship directly after New Year's. The estimated time required for inflation is four weeks which means that the ship would be ready to fly by the first of February. Whether or not the ship is ready by then remains to be seen.

On December 3, I wrote to Dr. Arnstein that "it looks as though Fickes should plan to be here about the middle of January, and not a great deal later if he is to see any of the inflation." In a letter to Mr. Litchfield on January 13, 1936, I advised that "inflation of the LZ 129 has been temporarily postponed but indications now are that the gas will be turned on in about a week. My guess is that the ship will not be ready to fly until March or the very end of February." Actually the first cell was installed in the *Hindenburg* on January 16 and gassing started, a second cell was installed on January 17 and "by tomorrow noon," I wrote to Scotty Peck,

> I expect that both cells will be about 50% inflated. The estimate for the time required for complete gassing is from a minimum of three weeks, if the weather is warm and everything goes well, to about five weeks. Four weeks should be a good approximation. If we allow another two weeks for the finishing touches and the shed tests the ship should be ready to fly in six weeks or the beginning of March. . . . So far everything has gone along very smoothly with the installation of the cells. The positioning, insertion of the axial corridor in the tunnel, and positioning of the corridor in the ship has all been accomplished without any unforeseen difficulties. That central corridor is a problem but it seems to be pretty well overcome. Of course this installation is much simpler now that the double cells are not being used.

Actually the first flight was made on March 4, 1936, with many details incomplete, the aim being to have the *Hindenburg* ready for a first flight to Rio, announced in advance, to commence at midnight March 30–31. As I wrote on February 28 to Mr. Litchfield, "all major shed tests have been postponed and the last coat of dope has yet to be applied to the outer cover. This will now be done between flights."

While I had expected Karl Fickes to arrive before the first flight, his plans were changed and it appeared I would still be the sole Goodyear representative in Friedrichshafen for some time.[2] Accordingly I asked Mr. Litchfield's permission to fly in the *Hindenburg* until she made her first voyage to the United States.

> The tentative program for the new ship is to make the first long flight to South America as a gesture of friendliness to the Brazilians and then to start flying to the States. The first long flights are of particular interest to me for it will be on these few flights that we will see how the ship stands up and where alterations will have to be made. The later flights during which a routine of operation has been established will also be extremely interesting but more so to Karl since his interest is more in operation.

Mr. Litchfield replied affirmatively on February 5, and I promptly approached Dr. Eckener to ask if he was agreeable to my participating in the flights of the *Hindenburg.* On February 28 I wrote to Mr. Litchfield:

> The situation is such that he feels that one Goodyear man can definitely be carried on all flights, and although at times there may be some question as to carrying a second man, the Doctor took the attitude that it could probably be arranged somehow. . . . The one exception to the above seems to be the first flight as the Doctor has already informed our Navy representative, Commander Peck, that the first flight is only for the crew and no observers will be carried. . . . It is proposed to fly the new ship nonstop to Rio, a distance of about 6000 miles, with a load of passengers. It is also contemplated to make the return flight,

> against prevailing winds, nonstop. Comparison of such a performance with respect to that of Pan Am's *China Clipper,*[3] which seems to have difficulty flying 2200 miles, let alone carrying any pay load, should really sell people on airships if the facts could only be presented to them.

As the first test flight or maiden flight of the *Hindenburg* approached there was the usual question as to who would be aboard on this special occasion. The first report was that there would be nobody aboard except Luftschiffbau personnel, but at the last minute an exception was made in my case, and I was the only one aboard who was not in the Luftschiffbau. It was a very special arrangement for me to be aboard, and I was thereafter on all the test flights. I believe this was the result of my close association with Knut and Dr. Eckener.

My station aboard the *Hindenburg* for takeoff and landing was the navigation room in the control car which was an area from which the entire landing and takeoff maneuver could easily be observed as well as all the activity in the forward section of the control car. It was an ideal spot. During the first flight as well as all later flights, I had no schedule of duties and was free to move about throughout the ship as I pleased. This tied in beautifully with my responsibility to Goodyear for I was to note anything and everything that took place aboard the ship and to be in a position to report on the ship and all of its inner workings. When things were quiet in the control car I would roam through the ship, literally from stem to stern, to all parts that were accessible, even into the power cars during the full power runs, and chat with the deck force concerning whatever duty they were performing. Whenever possible I would relieve the rudder or elevator man so he could take on other duties and I would have the opportunity to get the feel of the ship. It would be difficult to conceive of a more interesting arrangement.

At times I was quartered in the passenger quarters and at other times in the crew's quarters, depending on where space was available. My meals were taken either with the passengers or in the officers' mess, the latter being much more interesting for here I could get the officers' reactions to whatever was taking place. I also was provided firsthand with information that otherwise could not have been obtained. After almost two years of close association with the flight personnel, I found they were only too happy to discuss their ship and the various facets of the operation of the ship. I might as well have been one of the crew, except that I had no specific duties and was not on the Luftschiffbau Zeppelin payroll.

The first flight of the *Hindenburg* took place on March 4, 1936, and lasted 3 hours 6 minutes. Since Scotty Peck could not be on board, I gave him my camera so he could take photographs of the undocking and lift-off. What looks like trailing smoke in one photograph is dust and dirt being blown off the top outer cover by the air flow over the hull. The ship, about 90 percent full of hydrogen, was undocked in a light down-hangar wind of about 4½ MPH, only nine minutes being required from the start of undock-

The *Hindenburg* on her first flight, March 4, 1936. What looks like smoke is dust blowing off the outer cover of the ship. (Photo by Commander Scotty Peck for Harold G. Dick)

ing until the ship was in the air. For some time the ship cruised over the Bodensee (Lake Constance) with engine speeds of 1100 RPM, until it was definitely ascertained that everything was going well. Engine speeds were then increased to 1330 RPM which yielded an approximate (and evidently too high) speed of 77 MPH. As the ship behaved very well, preliminary turning tests were made at these two speeds, using only 10 degrees of rudder. During these tests, engine RPM, air speed, rudder angle, elevator angle, pitch angle, angle of attack (at frame 142.75 and about 5 feet 9 inches from outer cover) and altitude were recorded. The operators then ran a few tests to familiarize themselves with the ship, such as taking a few static weigh-offs with the engines idling, and reversing the engines at various speeds to determine how quickly the ship could be stopped. With all engines idling (this yielded an air speed of 27 MPH), it was found that the ship could be stopped in one and a half ship's lengths, which was considered better than could be done with the *Graf*.

The ship in general behaved very well and everyone was very enthusiastic about it. In steady air the ship was so stable that it almost flew itself, but at the same time it answered the rudders and elevators very well. The absence of vibration in the structure was very noticeable for the structure around the power plants was very quiet. The outer cover, on the other hand, was by no means as quiet as it might have been. I thought additional doping might help, but antiflutter wires and girders would have to be added in quite a few panels. The fins were very quiet, even quieter than those of the *Graf* with the after engine car shut down.

Practically nothing was done to the ship to get it ready for the second flight, which took place on the following day, March 5. This was devoted to speed trials at various RPMs and to a few more turning tests. When most of these were completed, the ship was flown to Munich and back to Friedrichshafen via Augsburg. The speed tests indicated that the LZ 129 had good speed, but that the methods of measuring air speed were not satisfactory. One air-speed meter of the impeller type, and another of the dynamic head type, were suspended on lines 82 feet long from frame 140. After all corrections were made, including one for proximity of the meters to the hull, the speed came out at 85 MPH, which was obtained at an average RPM of 1450 corresponding to 1030 HP per engine. During this full power run I climbed out into one of the forward power cars. The whole power plant at this high speed was particularly steady, even more so than at the lower speeds. The adjoining structure was also very quiet and vibration was noticeable by its absence.

The third flight on March 6 was made for the DVL[4] and a series of tests of air speed were made at various combinations of engines and engine speeds. The Luftschiffbau Zeppelin did not appear to be particularly interested in the tests which were made only to have the ship approved. Some damage was done to the control car in landing from this flight: the landing was made on the long lines and everything was well under control, though the ship was a bit heavy. With one and a half tons of emergency ballast forward it was decided to hold the water and to let the landing wheel take the shock. The rate of fall was indicated as 130 to 156 feet per minute (which was probably faster than intended) but as the stern struck first the actual speed was probably another 20 feet per minute higher at the control car owing to the whipping effect. The stern wheel took the shock all right. At the control car the results were not so good. Spectators on the ground reported that first the legs of the landing wheel compressed, then the wheel flattened out practically to the hub, allowing the fairing aft of the wheel to be forced into the ground, and then the whole assembly was forced up into the control car. The floor of the navigation room was pretty well stove in. The entire floor structure under the navigation room was cut out and had to be rebuilt. Three shifts of eight men each worked in order that the LZ 129 would be ready for the next flight.

The fourth flight was on March 17–18, of 22 hours 37 minutes duration. This, with an intermediate landing next morning followed by a flight of 7 hours 6 minutes, represented the final acceptance trials for the DVL. The takeoff was downwind from the west end of the building dock, with a wind of 7 MPH, and as it was thought best to play safe, the ship was made very light, approximately 3300 pounds as compared to about 1760 pounds for the average takeoff. Everything was being done "according to the book." Turning tests were made at different engine speeds. The rudder was applied as quickly as the rudder man was able to, and with all engines at 1425 RPM the rudder was swung from 15 degrees port to 15 degrees starboard as quickly as possible, but even so required several seconds. Preliminary figures indicated that the LZ 129 had a turning radius of about 4000 feet.

Pitching tests were made later in the day, but at lower speeds (approximately 67 MPH) for one engine was shut down at the time owing to a pump failure. The elevators were swung to 15 degrees up as quickly as possible and held there until the ship assumed an angle of pitch of 10 degrees. At this instant the elevators were swung to 15 degrees down as quickly as possible, and held there until the ship was at 10 degrees pitched down by the bows. When the ship assumed the latter angle she was again leveled off. The ship changed altitude 260 feet from the start of up elevator to down elevator. All these maneuvers were below pressure height. Although a lot of gas had been valved from the ship, she had not yet been taken through pressure height for she was heavy aft (and heavier than expected) and the officers did not wish to blow off any more gas from the after cells than was necessary.

The sixth flight, on March 23, was the first with passengers. About eighty newspapermen were on board. In preparation for the coming propaganda flight, loudspeakers had been installed to beam voice messages to the ground, but people on the ground said that the broadcasting could scarcely be heard above the engine noise. For the first time the LZ 129 returned not to the building shed at Friedrichshafen, but to the operating hangar at Löwenthal nearby. This was the first time that the new mast and trolleys had been used; the time from landing to being inside the dock was forty-nine minutes. About three hundred fifty men were in the ground crew at Löwenthal, forty of them pulling the new mooring mast!

The next two flights were of historic, even decisive significance, leading to Dr. Eckener's being declared a "nonperson" by the Nazis. His influence in the Luftschiffbau was undermined; that of Captain Lehmann was enhanced. Dr. Eckener's safety-first precepts were ignored with impunity, and it was clear that the growing control of the passenger Zeppelins by the Nazi regime posed a real threat to their future as transoceanic vehicles. In the Introduction I have told the story of the tail smash that occurred when Captain Lehmann insisted on starting the propaganda flight in the face of a fresh down-hangar wind in order to please the propaganda minister, Dr. Joseph Goebbels. In his book published in 1937, Captain Lehmann briefly dis-

The *Hindenburg* ready for the March 26, 1936, propaganda flight after the lower fin had been repaired. The *Graf Zeppelin* stands by overhead, since the two ships were to fly together over every town and city in Germany of 100,000 inhabitants or more. This is a rare photograph of the two giant airships together. (Photo by Harold G. Dick)

misses the accident while blaming it on the clumsiness of the ground crew.[5] Dr. Eckener was fully justified, however, in his criticisms: the 18 MPH wind was too strong to risk the difficult downwind takeoff, in which the airship, with engines idling, was made light aft so that the tail would ascend first and the ship would then be lifted dynamically by the wind blowing on her underside.

Such was the theory. In fact, as the stern rose, the elevator man, Kurt Schönherr, put on *up* elevator so that the wind could get underneath and lift the tail. At this point a gust, probably intensified by turbulence spilling over the downwind end of the hangar, slammed down the elevators—hinged surfaces each with a span of nearly 50 feet and an area of more than 700 square feet—with such force that the elevator wheel in the control car was torn out of the hands of the elevator man, throwing him across the car. The elevator control chains in the control car brought up with such force against the stops that the chains broke. Out of control, the big ship free ballooned across the city and out over the lake. Elevator control was quickly transferred to the emergency station in the fore part of the lower fin, and the elevator control chains were replaced. Schönherr, sent to a local hospital for examination, was found to have no broken bones, but his hands were badly bruised and swollen.

The propaganda flight itself, with fifty-nine officials as passengers, lasted 74 hours 14 minutes, slightly over three days, and covered 4150 miles. Dr. Eckener, as a sign of his displeasure, refused to participate. Instead, with continuing tension between him and his chauvinistic subordinate, Captain Lehmann, it was necessary to have a formal agreement—that if both men were aboard the *Hindenburg* one of them acted as commander, while the other would be simply a passenger and would play no part in the operation of the ship.

The March 29, 1936, referendum concerned annexation of the Rhineland. Broadcasting appeals to the electorate to vote "ja," playing stirring music such as "Deutschland über Alles" and the Nazi "Horst Wessel Lied," and dropping propaganda leaflets, the *Hindenburg* was seen over the major cities of the Reich from East Prussia to Oldenburg, from Swabia to Pomerania. Over Aachen a polling booth was rigged on the port promenade deck, and the 104 people aboard cast 104 "ja" votes for the Führer. None of the three Americans in Friedrichshafen—Scotty Peck, Karl Fickes, and myself—was aboard, word having come from Berlin that we were to stay home. The reason was quite obvious. Had we been aboard, we would have seen just about all of the military preparations being made in Germany at that time.

I would be compensated, however, by being included on the *Hindenburg*'s first ocean crossing, the previously advertised voyage to Rio, with the ship scheduled to depart early on the morning of March 31, 1936.

Chapter 12 *Transatlantic Flights in the* Hindenburg

he *Hindenburg*'s first crossing of the South Atlantic was a spectacular flight and an amazing undertaking considering the distance involved and the fact that the new ship had only 128 hours 8 minutes in the air. The Germans had all the confidence in the world, for they felt they had a good ship in the *Hindenburg* and they were sure the ship had good engines. The *Hindenburg* was still hangared at Friedrichshafen, the new overseas base at Frankfurt am Main not being completed yet, so the flight was to be nonstop from Friedrichshafen to Rio. Captain Lehmann was in command, with General Christiansen, also a director of the Reederei, playing a conspicuous role with the guests.[1] Dr. Eckener as a passenger was in the background.

Takeoff was on March 31, 1936. Our normal route would have been from Friedrichshafen across Lake Constance to Basel, over France and down the Rhone valley to the Mediterranean, then through the Straits of Gibraltar, along the coast of Africa, over the Cape Verdes, and across the South Atlantic to Rio. (Unlike the *Graf*, the bigger *Hindenburg* did not have to make an intermediate stop at Recife.) This was the shortest and most direct route. The French, though having no great love for the Germans, had allowed the *Graf* to follow this route. But now they would not permit the *Hindenburg* to fly over France. Thus, still making the flight on schedule, we flew from Friedrichshafen in southern Germany north to Holland, out over the English Channel, around the northwest corner of France, then south past Spain and Portugal, over the Canaries and Cape Verdes, and across the South Atlantic to Rio, a total distance of 5942 miles, and a flight duration of 100 hours 40 minutes.

Although this was a much longer distance than any airplane flight made up to that time, we carried thirty-seven passengers, fifty-four men in the crew,[2] a spare Daimler diesel engine weighing about 2 tons,[3] 1.3 tons of mail and freight, all the provisions for the return flight, and 14.3 tons of water ballast—and on arrival at Rio we had sufficient fuel aboard for a further 1040 miles. The loads actually carried are shown on this chart:

Loading (Preliminary) for Flight #9, LZ 129
Friedrichshafen to Rio, on March 31, 1936

Passengers (37)	2,960 kg
Engine personnel (23)	1,840
Deck force (21)	1,680
Steward personnel (10)	800
Baggage for passengers	600
Baggage for crew	1,080

Provisions	3,000
Freight	1,269
Mail	84
Fuel oil	55,230
Lub. oil	4,000
Kerosine	—
Reserve parts	1,200
Reserve radiator water	1,400
Drinking water and liquors	1,150
Trim ballast	11,300
Emergency ballast	3,000
Miscellaneous	1,120
Bedding, utensils, etc.	2,500
Moisture	1,000
Lightness	1,000
Total	96,213 kg
Dead weight (approx.)	118,000
Total	214,213 kg
Required gas volume	202,000 cbm

Takeoff, Friedrichshafen, 0432 gmt, March 31, 1936
Landing, Rio, 0912 gmt, April 4, 1936

Total time	100:40
Total distance	5948 nautical miles

Fuel oil required	46,530 kg
Lub. oil required	1,700 kg

$b = 729$ mm
$T_1 = 9.0$C
$T_g = 13.0$C
$R_f = 85\%$

sH = 0.090
C = 1.0935 (?)
1.070*

*Specific gravity of hydrogen

Weather conditions were not severe and in general the weather was quite good. The air was a bit bumpy over the Bay of Biscay where we also ran into head winds.

The passengers seemed well pleased with the accommodations for they had plenty of room to move about, the visibility at our low altitude—650 to 800 feet—was tremendous, and the food prepared by the chef from the Kurgarten Hotel, the finest hotel in southern Germany at the time, was easily the equivalent of that provided by the luxury liners of that period. All of the food was freshly cooked and the Germans had their freshly baked hot rolls each morning for breakfast.

During the first day the ship flew over Stuttgart, Frankfurt, Cologne,

Chart showing the *Hindenburg*'s first South American flight: Friedrichshafen to Rio de Janeiro and return, March 31–April 10, 1936. (From the collection of Harold G. Dick)

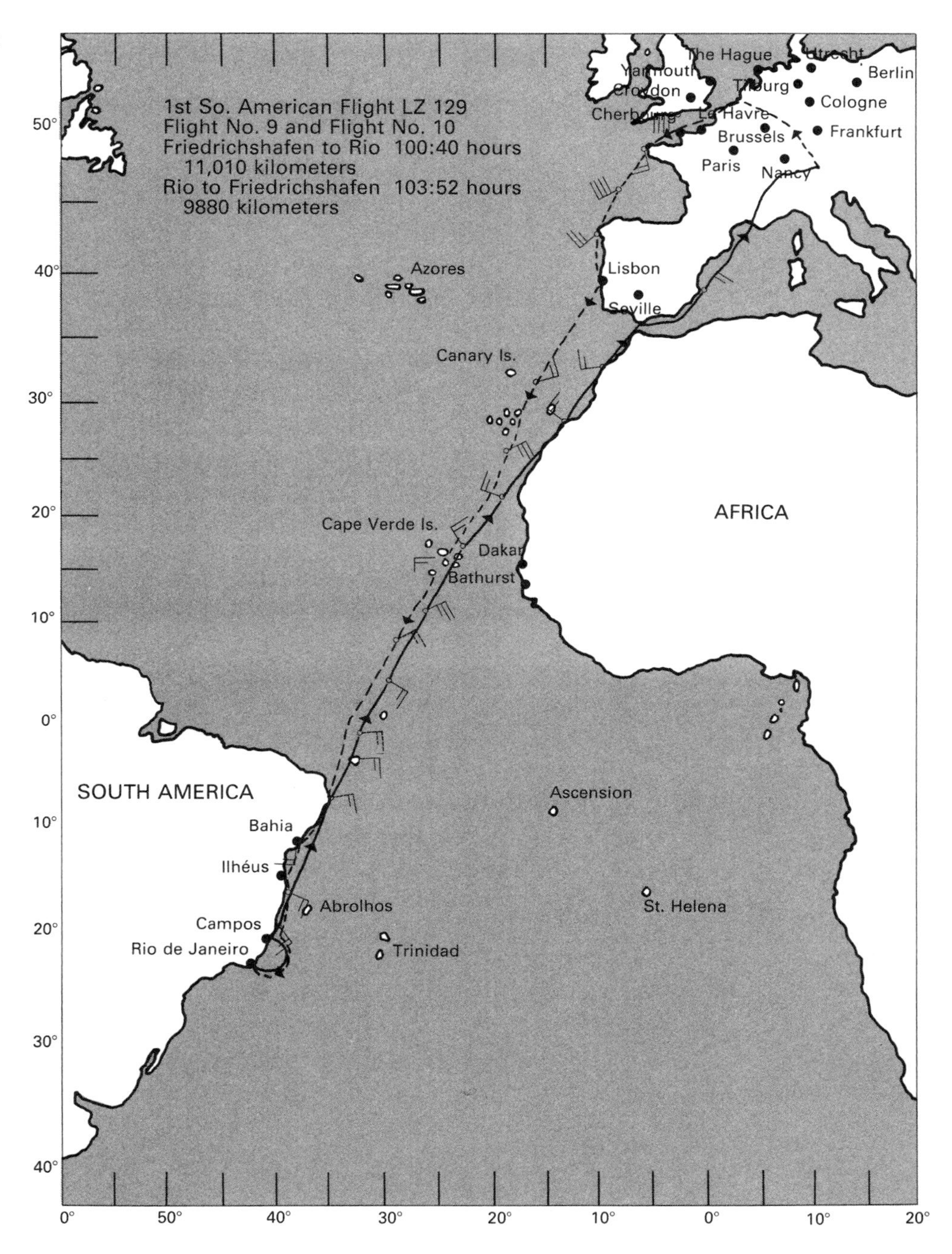

Tilburg, Dover, and past Dungeness and the Île Vierge on the north side of the Brittany peninsula near Brest. The next morning the ship was at Cape Villano on the northwest coast of Spain, and by late evening was at Las Palmas in the Canaries. The following forenoon the *Hindenburg* passed Boa Vista and Porto Praia in the Cape Verdes. Early next day she was at Fernando de Noronha and a few hours later at Pernambuco. By nightfall the *Hinden-*

burg was at Bahia and the next morning arrived at Rio de Janeiro.

The flight was longer than anticipated because of the longer route around France but the passengers did not object for they arrived rested and six days earlier than if they had traveled to Rio by ship. The fare to Rio was two thousand marks or five hundred dollars at the existing rate of exchange. This was comparable to the charge for the best ocean-going vessels at that time.

The flight itself was relatively uneventful. The landing in Rio was not so uneventful. A new hangar had been built at Santa Cruz, about forty miles southeast of Rio, and it was supposed to have been ready for the ship. A ground crew of two hundred and forty men was on hand but most of them stood around looking at the ship and holding the handling lines without evidently knowing what to do with them. The entire field was under about six inches of water from heavy rains. Because of the difficulty in handling the ship in these conditions, it was decided to keep the stern in the air until the ship could be put on the mast. This resulted in getting the nose cone below the cup on the mast and shearing off the mooring cable, making it impos-

The new hangar at Santa Cruz, south of Rio. When first used by the *Hindenburg* on its first flight to Rio, the airfield was under six inches of water.

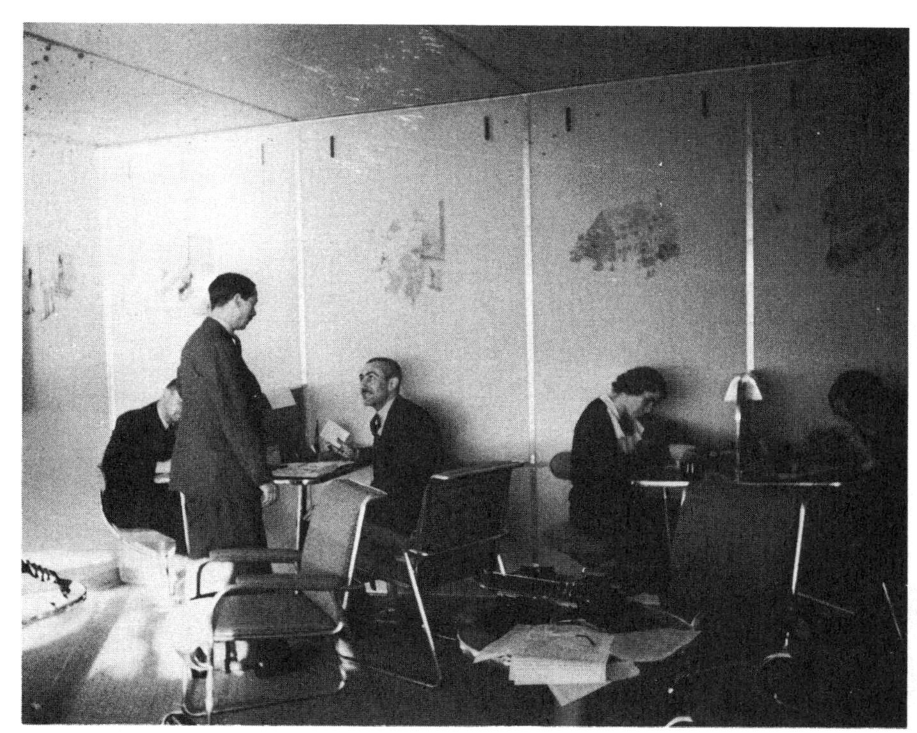

TOP LEFT: Aboard the *Hindenburg* on its first flight to Rio, March 31, 1936. Harold G. Dick, *back to camera,* is conversing at dinner with Knut Eckener.

TOP RIGHT: Aboard the *Hindenburg* en route to Rio: passengers in the writing room. (Photo by Harold G. Dick)

Harold G. Dick in the navigation room of the *Hindenburg* on its first flight to Rio. (From the collection of Harold G. Dick)

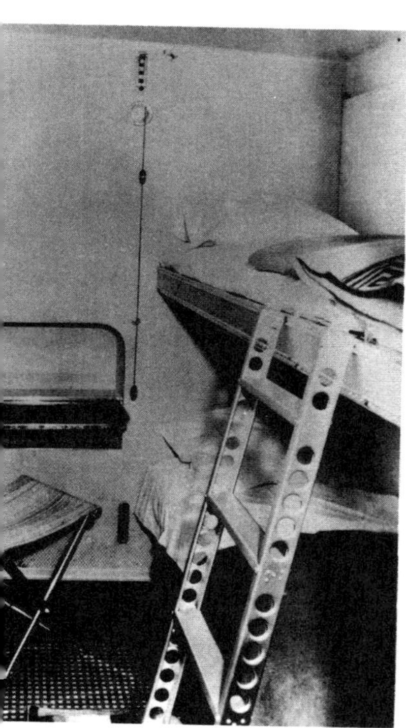

ON PAGE OPPOSITE, TOP: Aboard the *Hindenburg* en route to Rio. Passengers in the starboard promenade, view looking forward. To the left is the lounge and forward to the left the writing room. Three windows are open. (Photo by Harold G. Dick)

BOTTOM LEFT: Passenger cabin in the *Hindenburg*, with upper and lower berths, call bell (rope on wall), vanity table, tip-up washbasin (*at left*).

BOTTOM RIGHT: Lounge with map of the world on interior wall. Writing room is in the background.

sible to use the mast. The ship was then walked, with the help of the engines, up to the hangar doors until the bow was almost inside the hangar. The forward trolleys were then secured, the stern hauled up, and the ship walked into the hangar. The entire maneuver took about an hour and there were times when it seemed that the safest place would have been on the ground, even in six inches of water.

While over the South Atlantic, Dr. Eckener had come to me and asked if I could help him frame an answer to a radiogram he had received from Reuters in London. The radiogram stated that it was reported that the Nazis were ostracizing Dr. Eckener and he would not be allowed to return to Germany, apparently because of statements he had made concerning the Nazis. As we later learned, the Doctor's comment about the use of airships for political purposes being the end of the airship had been reported to Berlin. Since the Doctor had nothing official from Berlin we could only profess ignorance in our reply. I did feel highly complimented that the Doctor would ask me to help him in such a personal matter.

As it turned out, Dr. Eckener was allowed to return to Germany but the Nazis took his citizenship away from him. During this period, which was about nine months, he could come and go as usual but he could not be quoted or given public recognition of any kind.

The return flight to Friedrichshafen was far from uneventful but was not as dire as the news reports may have indicated. Actually it was a very successful flight, leaving on schedule with thirty-eight passengers aboard, covering 5347 miles nonstop, with an elapsed time of 103 hours 52 minutes. (On arrival in Friedrichshafen, the ship had 50,450 pounds of reserve fuel aboard or sufficient for another forty-five to fifty hours at normal cruising speed.)

Takeoff for the return flight was on April 6 in the morning in calm weather with a ground crew of only two hundred. The mooring cable had been repaired and the mast was used during the undocking procedure. With a morning takeoff the passengers had the opportunity to see all the coastal sights, for the ship cruised along the Brazilian coast as far as Pernambuco at an altitude of 650 to 800 feet. Normal cruising altitude for most of the flight was 650 feet, but on approaching the African coast the ship was taken up to 6400 feet in order to obtain more favorable winds. On one occasion, at an altitude of 4750 feet, the ship encountered the warm dry air coming off the Sahara Desert with humidity as low as 6½ percent.

Weather conditions in general were quite good. Some old thunderstorms were encountered coming out of the doldrums. At this time the air was rather turbulent and one gust took the ship up 525 feet at a rate of 880 feet per minute, which was not excessive. In passing through these old thunderstorm formations, our altitude was 800 feet. During the entire flight eight tons of rain water were collected from showers.

About noon on the first day out of Rio, the ship was at 800 feet; this was just about pressure height with all gas cells 100 percent full. The passengers were having their noonday meal, and the watch officer on duty was Knut Eckener, with Heinrich Bauer as navigator. The automatic valve in cell number 3 in the stern opened—and for some reason popped wide open and stuck there. The valve could not be closed until the gas level reached the valve and by that time about 160,000 cubic feet of gas had been lost and the cell was half empty. Eighteen degrees of down elevator was required to hold the ship on an even keel; speed was 74 MPH. Sixteen men were sent up into the bow, one tank of fuel was pumped forward, the two aft emergency tanks were connected into the lines allowing the water to run to the forward tanks, all cells were connected into the inflation line to bleed gas back into cell number 3, and although this all took place in a few minutes the ship was held on an even keel. Even Dr. Eckener, who was with the passengers at this time, did not notice that anything out of the ordinary had taken place. Losing 160,000 cubic feet of gas was roughly equivalent to losing 4½ tons of lift in the after section of the ship.

Why the automatic valve on cell number 3 stuck wide open was never determined, although nothing of the kind ever happened again aboard the *Hindenburg*. Because of the location in the tail of the gas shaft serving cells 2 and 3, the shaft did not terminate in a conventional hood on top of the ship but in an opening on the side of the upper fin. The opening was flush to the fin surface, with louvres, and there was the suspicion that the louvres had something to do with the difficulty. Henceforth the automatic valves on cells 2 and 3 were not trusted, as observation showed they vibrated when they started to open. Instead, whenever the ship approached pressure height, the hand-operated maneuvering valves were used to prevent a repetition of the mishap that occurred on the flight home from Rio.

A minor problem developed with the elevator control wheel but this was not serious. Both the rudder and the elevator control wheels could be shifted from manual to servo control by means of a clutch. The clutch on the elevator wheel jammed making it impossible to shift from servo to manual. While the clutch was being released, which took about an hour, the elevator control was shifted to the emergency stand in the lower fin. For the remainder of the flight only manual control was used on the elevator wheel in the control car.

As far as the passenger quarters were concerned, it was found that somewhat better ventilation was desirable. This was easily remedied when the ship returned to Friedrichshafen.

The problem of the stuck valve had been attended to so quickly and expertly that none of the passengers, and particularly none of the numerous reporters on board, had any inkling of the malfunction. The same was not true of the major problem of the homeward voyage, which was with the engines. The passengers and newsmen aboard were quite concerned because, from the passenger quarters, they could at times see that the two forward engines, number 3 and number 4, were not operating.[4] Some of the reports they sent home were sensational, and made it appear that the *Hin-*

denburg had narrowly escaped disaster off the African coast. Although the situation was serious, at no time was the ship in any immediate danger.

All four engines were in operation until the ship was about nine hours beyond Fernando de Noronha and out over the South Atlantic. One engine was shut down briefly for twenty-four minutes and then the ship continued with all four engines for another six hours. For four hours the ship operated with three engines and then for two hours with only two engines. After number 2 engine was repaired, the ship ran on three engines for seven hours and then all four engines were again operating. Twelve hours later when the ship was off Cape Juby, number 2 engine was stopped and the flight was completed on three engines.

By the time the ship had reached Gibraltar a decision had to be made concerning the remainder of the flight. Head winds were reported over the English Channel and because the ship's speed with three engines was 62.6 MPH, it was decided to proceed across the Mediterranean and request permission to fly over France via the normal route up the Rhone valley. If permission were not granted the alternative was to fly over the Alps, a route that would take the ship over part of Italy and Switzerland. This was not a desirable route, as the weather over the mountains was cloudy and visibility was poor, but it was nonetheless a feasible alternative. The French in fact did grant permission for the ship to proceed up the Rhone valley. This seemed a very logical decision since they were granting permission for the *Graf Zeppelin* to use this route and had done so for several years.

Since the Germans particularly during the Nazi period refused to admit failure in enterprises involving national prestige, no statement was made by the Luftschiffbau Zeppelin concerning the engine failures on the *Hindenburg*'s homeward flight from Rio. Historians, I understand, have difficulty obtaining information on this serious matter, and I have seen a variety of explanations in print. Even Karl Fickes and Scotty Peck had somewhat different impressions from mine. Crew members felt free to talk to me and I felt that the report I sent home to Dr. Arnstein on April 13, 1936, was factual and accurate:

On the outward journey to Rio practically no difficulty was experienced and the engines were operating at various RPMs between 1200 and 1425 depending on conditions.

In Rio, three pistons were pulled from one of the engines as there were indications that the rings had carboned up because of running at low speed during the three-day propaganda flight, and the engine was burning oil. Although a fifth Daimler was carried in the keel, there was no thought of replacing the engine with the spare one.

Upon leaving Rio all the engines were operating satisfactorily except that this one engine was turning up at lower speeds while the pistons were being run in. Air speed was 69 MPH, which was rather low but yielded very low fuel consumption.

Thirty-eight hours out of Rio, number 4 engine was shut down with a damaged piston and cylinder. The wrist pin, as was later determined, was of faulty material and failed, causing damage to the piston and the cylinder. Work was immediately started to repair the damage; this took sixteen hours. At the end of this time the engine was run again. All moving parts had been removed from the damaged cylinder and the engine was run on fifteen cylinders. It continued to operate this way throughout the remainder of the flight, a further fifty hours. The operation was not particularly satisfactory for towards the end of the flight the engine was knocking quite badly.

Four hours after number 4 had failed—that is, forty-two hours out of Rio—number 2 engine was shut down. A bearing cap bolt had failed and the cap had broken loose and fallen into the crank case. The cap was that on the big end of the connecting rod and in this particular engine there were two of these caps since the rod was split. Two hours were required to remove the cap from the case and then the engine was run again but with only one bearing cap to take the load instead of two. For twenty-five hours the engine ran without trouble, but then the other cap let go and the engine had to be shut down again. This occurred sixty-nine hours out of Rio and the engine, being the test stand engine, had 968 hours on it at the time.

In order to have power available in an emergency, the engine was torn down and all the parts removed from the throw of the crank shaft to which the bad connecting rod had been attached. This required about sixteen hours and the engine could then have been run again, but on only fourteen cylinders. (In case of real emergency this type of engine could be run on only eight cylinders.) As the ship was not in any danger and in order to prevent any further damage to the engine—perhaps from torsional vibrations—the engine, number 2, was not run again for the remainder of the flight.

Engines number 1 and number 3 were reduced to 1200 RPM, probably to play safe, but also to keep from having all the thrust on one side of the ship. Air speed with two engines was 56 MPH. Engines number 1 and number 3 were not as smooth as could be, owing to irregularities with the injectors. The rings were pretty well carboned up and the engines had slight knocks.

For the landing maneuver, engines number 1 and number 3 were used to stop the ship, while engine number 4 was only idling. The time to stop the ship was appreciably longer than normal. The landing itself, however, was skillfully executed and very satisfactory.

The failure of the bearing cap bolt was attributed to faulty design, and bolts of a new design were installed in all the engines. The old design was such that fatigue failure was possible, for the threads had been cut with a sharp root and the bolt had been drilled out, reducing the cross section beneath the thread. The new bolt had a round root and was not drilled out. A comparison of the two designs shows that the new bolt was of much sturdier construction.

Dr. Eckener was deeply concerned about the engine failures on the homeward flight from Rio, because they led to unfavorable publicity for the *Hindenburg* and had needlessly worried the passengers. Privately he blamed

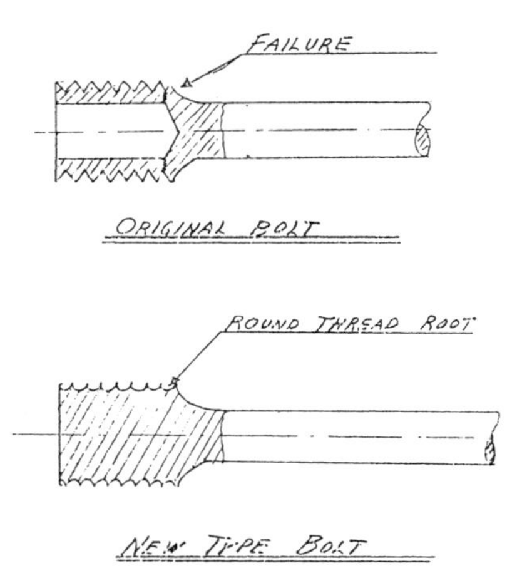

Comparison of the old and new bolts in the *Hindenburg* engines. (Drawing by Harold G. Dick)

the fiasco on Dr. Goebbels, whose demand that the new ship make the three-day propaganda flight from March 26 through 29 had left no time for the twelve-hour full power trial that he would have insisted on before setting off for South America with passengers. Nor would he have hesitated to cancel the announced March 31 departure for Rio, regardless of considerations of prestige, had serious defects appeared. But the Reederei was now being directed by Captain Lehmann and his Nazi colleague, General Christiansen.

For a month the *Hindenburg* lay in the hangar at Löwenthal. The engines were returned to the Daimler works at Stuttgart for overhaul and cleaning up. Scoops were installed forward for the passenger quarters and ducts led up over the top of the cabins to give additional ventilation. An outlet was added to the ceiling of each cabin. The lower fin and rudder were faired in a bit more smoothly, the lower fin being cut away as far forward as the first intermediate forward of frame 20. The rudder was cut back to the lower fin support. The outer cover was retensioned aft of frame 20 and the entire cover was cleaned up and given its last coat of dope. A rain drip was added over the windows in the passenger quarters.

Finally, on May 4, 1936, the *Hindenburg* was brought out for a test flight of 7 hours 32 minutes to test the engines and to determine whether all parts of the ship were functioning properly. The test flight was satisfactory, and on May 6 the *Hindenburg* commenced her first flight to Lakehurst, with fifty passengers. Captain Lehmann was again in command, with Dr. Eckener aboard as a passenger.

The ship was undocked from the Löwenthal shed in light northeast winds and the course was laid over Stuttgart, Frankfurt, Cologne, across Holland, and down the English Channel. Weather conditions did not appear favorable for a fast flight for there was a low centered over the North Atlantic and a high pressure area centered in the vicinity of the Azores. The choice was of taking a southerly course through the center of the high or of attempting to take advantage of the spreading lows (which would mean flying a northerly course). A northerly course was selected, and was altered from time to time to take advantage of troughs and secondary lows when they were indicated. It was noticeable that there was a considerable lack of weather information over the Atlantic north of the shipping lanes. Such information would be of particular advantage to the airship.

No particular turbulence was encountered and there were no fronts deserving of the name. Conditions were much more favorable than those ordinarily encountered on the South American run to Rio. On the second night out, the air was not particularly smooth and short, sharp gusts of very short duration were encountered. The ship could be felt to shudder but the gusts were so brief that the elevator man did not even have to compensate for them. Although there may have been stresses set up in the ship, there was no difficulty at all in handling her.

Normal cruising altitude was 650 feet, but this varied at times in hopes of

The *Hindenburg* in flight: the lower fin and rudder have been reworked after the March 26 propaganda flight. The ground clearance of the lower rudder was increased from 8 to 14 degrees.

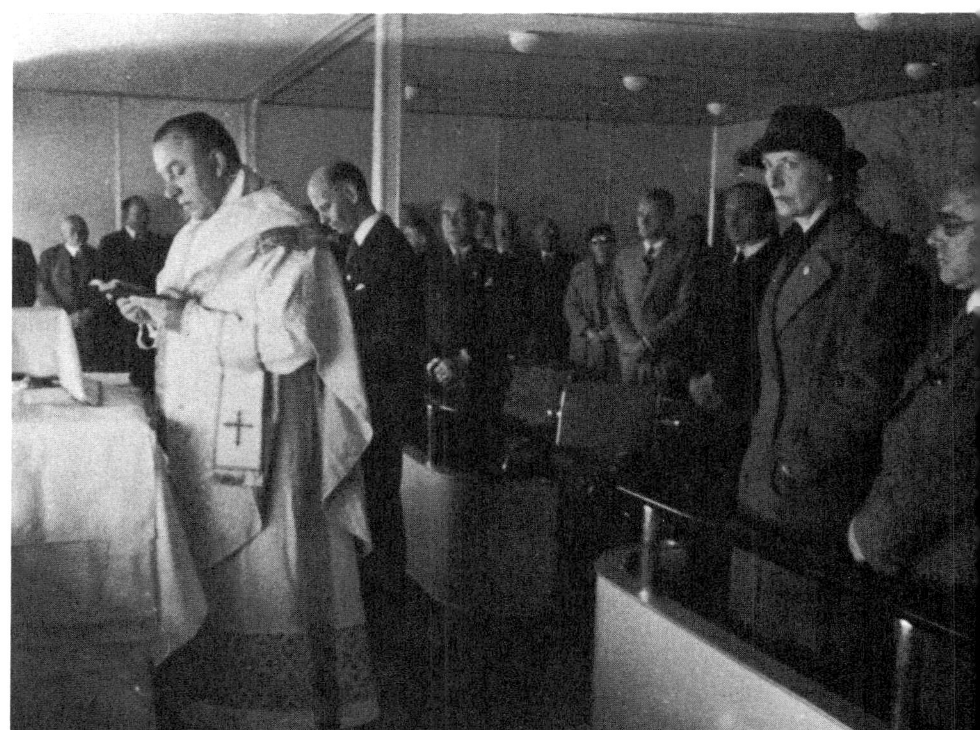

Father Schulte celebrating Mass over the middle of the North Atlantic on Sunday, May 8, 1936, during the *Hindenburg*'s first North American flight: Friedrichshafen/Löwenthal to Lakehurst, New Jersey. (Photo by Harold G. Dick)

picking up more favorable winds. The lowest altitude flown was 330 feet when it could be seen that the surface winds were more favorable than the winds at 500 to 650 feet. The ship was not flown for any length of time at 330 feet, but was flown considerably at 425 feet and a great deal at 590 feet. The flight time was 61 hours 41 minutes, and the distance 3805 miles from Friedrichshafen-Löwenthal to Lakehurst.

The engines on this flight functioned perfectly, the normal RPM being 1350, which varied from 1340 to 1360. During the entire flight there was no forced shutdown of any engine, not for one minute. The only engine shutdown at all was for Scotty Peck and Sir Hubert Wilkins to enter an engine car and make a photograph. Everybody was very well pleased with the engine performance on this flight.

The fuel consumption figures at the time of this flight, which contained approximations due to temperature changes but were used as the basis for loading fuel, are given below.

RPM	CONSUMPTION LB / HR	MPH
1450	1430	82.7
1350	1166	78.3
1250	880	69.3
1150	660	66.0

Oil consumption was given as 8.8 pounds per hour per engine. The switches on the servo motor control again did not appear satisfactory; they had failed before and there was another failure on this flight. The statoscope also failed to function properly.[5]

Beyond this the ship functioned perfectly throughout and there were practically no complaints to make. It appeared that there was too much of a head on the water lines to the cabins so that the faucets were not as satisfactory as they might have been. The heating system was capable of delivering sufficient heat, but better control by the stewards or an automatic control was desirable.

The landing at Lakehurst was carried out with about one hundred and twenty men in the ground crew, and for the first time the ship was landed on the short lines. The landing was very well executed and considering the small ground crew the entire maneuver went very well. The only difficulty was in putting the ship on the mast, this being made increasingly difficult by the fact that the control car had to be supported on crutches by the ground crew.

The lifesaving equipment consisted of four boats, carried in the two stub keels of the forward engine cars, each boat capable of carrying twenty-five people and inflating automatically when thrown into the water. In addition there were life jackets for each of the passengers. These life jackets, which were of the automatic type that inflate when immersed in water, were carried in boxes under the seats in the public rooms. Each of the crew was provided with a pillow that would keep him afloat.

I had the pleasure on this flight of renewing my acquaintance with Father Paul Schulte, the "flying missionary," whom I had first met in 1934. During the last year of World War I, Father Schulte had served as a pilot in the German air force. Sickened by all the killing, he had entered the Catholic priesthood after the war was over and in 1934 was a missionary in Africa. To bring medical aid to the natives was quite a problem, for transportation was by automobile and there were no roads to speak of. The result was that by the time help arrived for a seriously ill native it was usually too late. It was evident that if the natives were to receive effective assistance, a better means of transportation would have to be provided. Having been a flier himself, Father Schulte realized the airplane was the answer to the problem of faster transportation in Africa.

We became acquainted in Friedrichshafen. Father Schulte had visited the Pope, and the Vatican had authorized the purchase of two airplanes from the Dornier works at Manzell. At this time Father Schulte was about to take delivery of the airplanes.

He had also become well acquainted with Knut Eckener. When Knut organized a small group of perhaps ten of us for a *Bierabend* in Lindau, we took Father Schulte along and found he was delightful company. He was a big man with a tremendous energy and enjoyed the dark Bavarian beer, the pig shanks and sauerkraut, as much as any of us. During his stay in Friedrichshafen, Father Schulte joined us more than once on our outings.

Father Schulte soon took delivery of his airplanes and we heard nothing more of him until we were about to make the first North Atlantic flight of the *Hindenburg* to Lakehurst. He then showed up in Friedrichshafen and we learned that he was proceeding to the United States and from there to northern Canada where he was to do missionary work with the Eskimos. And he was of course going to use airplanes in his work. Being a bit of a promoter, Father Schulte had made arrangements to fly to Lakehurst with us on the *Hindenburg*. And since we were to be over the middle of the North Atlantic on Sunday, May 8, he planned to hold the first airborne Catholic mass on that day. He had all the necessary vestments with him, and the mass was attended not only by all the passengers, but also by those crew members who were not on duty at that time. Only one detail could not be supplied: the candles on the altar were not lit.

The return flight to Germany departed from Lakehurst at 3:28 A.M. on May 11, and was just as uneventful as the journey to the States had been. The weather was so good that it might be said the ship had gone out for a Sunday school picnic.

With a very flat pressure gradient along the eastern seaboard, light winds were experienced until the *Hindenburg* was well out to sea when following winds were encountered as the ship passed across the northern side of a high pressure area. As there was also a low pressure area concentrated in the northeast sector of the Atlantic, the following winds extended all the way to the Continent. Once again the reputedly stormy North Atlantic yielded high winds but no turbulence; there was actually no turbulence in the air at all except that due to convection and even that was very little. The ship was at all times very steady and more stable air conditions could not be desired.

Considerable fog was encountered off the Grand Banks and navigation made rather difficult, but occasional drift readings were obtained and after leaving the Banks the ship's position was corrected by about fifteen miles from the position of a small ship we encountered. From this point on, all navigation was by dead reckoning with the course set on the southern tip of Ireland. A landfall was made twelve miles north of the point on which the course was laid, but this was not entirely error for the course was laid in such a way that if there were any errors they would all be on the northern side in order that the ship would not pass south of Ireland. The landfall was also made within ten minutes of the predicted time.

For the first time, the *Hindenburg* landed at the new international airship base at Frankfurt am Main, where one hangar was finally ready. An advantage of operating from Frankfurt was that since it was only 300 feet above sea level, the *Hindenburg* could lift 13,500 pounds more than in a takeoff at Friedrichshafen, which lay 1000 feet higher. The flight time from Lakehurst to Frankfurt on this flight was only 49 hours 3 minutes, the distance 3597 miles.

The engines again functioned perfectly throughout the entire flight, without a single shutdown. Number 4 engine, however, showed indications of using too much oil and it was believed that the rings were stuck. This was verified after the cylinders were pulled. Number 2 engine had to be run at reduced RPM, 1300 instead of the normal 1350, since the car had become unsteady. The suspension was checked when the ship was in the hangar and the propeller was changed. It was thought that one of the blades might have absorbed more moisture than the others and been thrown out of balance.

Considerable rain water was collected, ten tons in all being added to the ballast aboard. This saved the valving of 305,000 cubic feet of hydrogen. The idea that rain water collection was not satisfactory (because at the time the ship could pick up water it would be too heavy) was obviously an error. As it was attempted to fly the ship between one and two degrees light when it was necessary to fly light, the ship always entered a rain squall light enough so that it could pick up water for ballast and in addition carry the increased weight of the wet outer cover.

For many of the *Hindenburg*'s crew, this flight was their first visit to Lakehurst, the main base of the U.S. Navy's lighter than air operations. Even for the oldtimers there was much at Lakehurst that was new, as the *Graf*'s last visit there had been in June 1930. While the *Graf* had then been brought into the hangar on a low traveling mooring mast mounted on crawler feet, the traveling mast concept had been developed further for the giant *Akron* and *Macon* airships into a massive affair rolling on railroad tracks 64 feet apart, while at the other end the ship's stern and lower fin were made fast to a ponderous "stern beam" 186 feet broad, which also rolled on railroad tracks. Out on the field the stern beam was to be transferred from the rails running straight out from the hangar to the "hauling up circle" on the field.

For this first North Atlantic crossing, it had been planned to take the *Hindenburg* into the hangar. Since it was my understanding that the decommissioned *Los Angeles* occupied part of the shed, I wrote several times to Lieutenant Commander George H. ("Shortie") Mills for the dimensions of the Lakehurst hangar and just how much space the *Los Angeles* occupied, sharing all this information with Herr Kolb, who was in charge of ground handling. In fact, there was no space problem when the *Hindenburg* was put in the hangar, since the *Los Angeles*, inflated but not in commission, was out on the field for mooring trials.

My German friends were impressed with the traveling mooring mast and felt it would be fine to have one like it at Frankfurt, but also agreed that it was too big an investment for a commercial company. Furthermore, it took twenty minutes to put the *Hindenburg* on the mast before leaving the hangar, causing concern about how much longer it would take to put her on the mast in gusty weather, after landing on the ground first as was the German practice. The stern beam was not satisfactory, particularly because of the time required to transfer the beam from the docking rails to those at the hauling up circle, and vice versa. Also, with the ship attached to the stern beam at only two points, the maximum crosswind that could be tolerated for docking was only 5½ mph.

Considering the short period that the *Hindenburg* lay over at Lakehurst—roughly a day and a half—it was decided not to put her in the hangar, which was time consuming, but to leave her on the mast at the riding out circle, with the lower fin resting on a riding out car that rolled on rails around the mast.

I was also aboard for the second North American flight of the *Hindenburg*, from Frankfurt to Lakehurst. This flight commenced at 4:33 a.m. on May 17, 1936, and owing to weather conditions took 78 hours 30 minutes to cover 3920 miles. This in fact was one of the longest flights to Lakehurst of the entire 1936 season. Shown below is the loading chart for this flight:

Loading for Flight No. 14, LZ 129
Frankfurt to Lakehurst, May 17 to 20, 1936

Passengers (41)	3280 kg
Engine personnel (21)	1680
Deck force (21)	1680
Steward force (11)	880
Baggage for passengers	——
Baggage for crew	1500
Provisions and liquors	——
Freight	——

Mail	130
Fuel oil	54160
Lub. oil	3400
Kerosine	50
Reserve parts	3500
Res. radiator water	150
Fresh water	7500
Trim ballast	4750
Emerg. ballast	2000
Miscellaneous	530
Fixed miscellaneous	——
Moisture	——
Lightness	——

Fuel consumed—	41,110 kg	Duration—	78:30 hrs
Oil consumed—	1,305 kg	Distance—	7238 km
			3920 nautical miles
Reserve fuel—	13,050 kg		
(At cruising speed)	24.6 hrs		

The takeoff at Frankfurt was normal and the ship proceeded via Cologne to the English Channel. Light winds from the southeast to northwest were encountered during the day but at evening, at about 6 P.M., a front of some intensity was encountered. Winds shifted from southwest, force 3 (8–11 MPH) to northeast, force 9 (41–48 MPH) with no change in temperature. Through the night the ship made good speed over the ground, 81 to 83 MPH, flying at 820 feet. By morning the winds had diminished and gradually shifted into the southeast. The ship at this time was just north of the Azores as the decision had been made to take a southerly route because of adverse winds to the north.

Throughout the day the winds became more southerly and increased in intensity as the ship approached a front that was lying almost north to south. For a period of about seven hours the wind swung more to the south and varied from force 8 (34–40 MPH) to force 10 (49–56 MPH). As the wind became stronger altitude was reduced from 820 feet to 500 feet.

The front was encountered at 1400 Greenwich Mean Time with the wind directly out of the south at force 9 (41–48 MPH). The seas below were running so high that it seemed the *Hindenburg* was almost on the surface. Estimates were that the sea must have been at least 50 feet high. With the front running north to south, the ship took a southerly course following the front and looking for a light spot where the ship could be taken through or under the front. During this our ground speed dropped to as low as 30 MPH and for three hours we made very little progress.

At 1700 GMT, with the ship again at 820 feet, temperature 62° F, wind south-southwest force 9, it appeared that there was a light spot in the front

OVERLEAF: Chart showing the *Hindenburg*'s second North American flight: Frankfurt, Germany, to Lakehurst, New Jersey, May 17–20, 1936. (From the collection of Harold G. Dick)

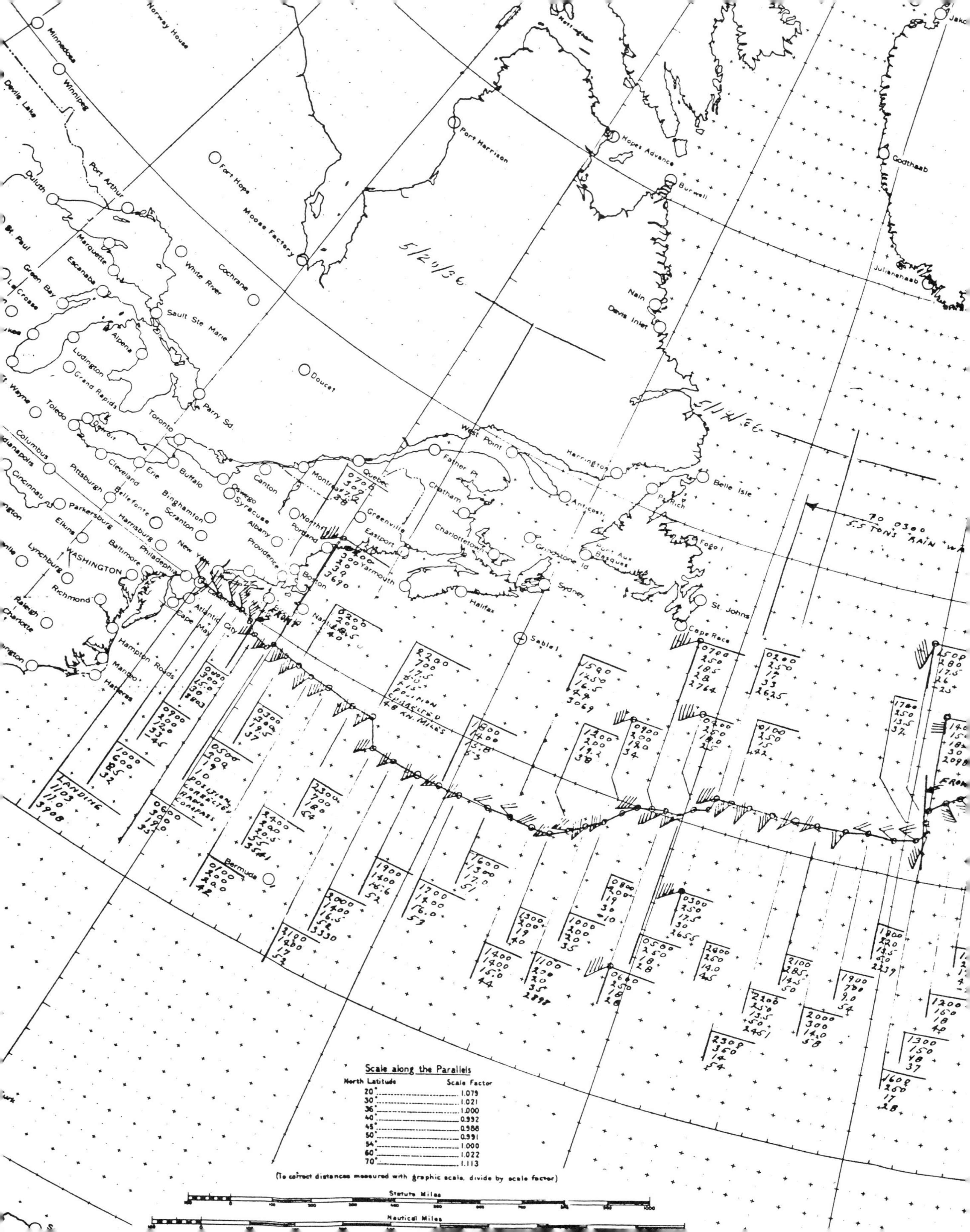

Scale along the Parallels
North Latitude Scale Factor
20° 1.079
30° 1.021
36° 1.000
40° 0.992
45° 0.986
50° 0.991
54° 1.000
60° 1.022
70° 1.113
(To correct distances measured with graphic scale, divide by scale factor)
Statute Miles
Nautical Miles

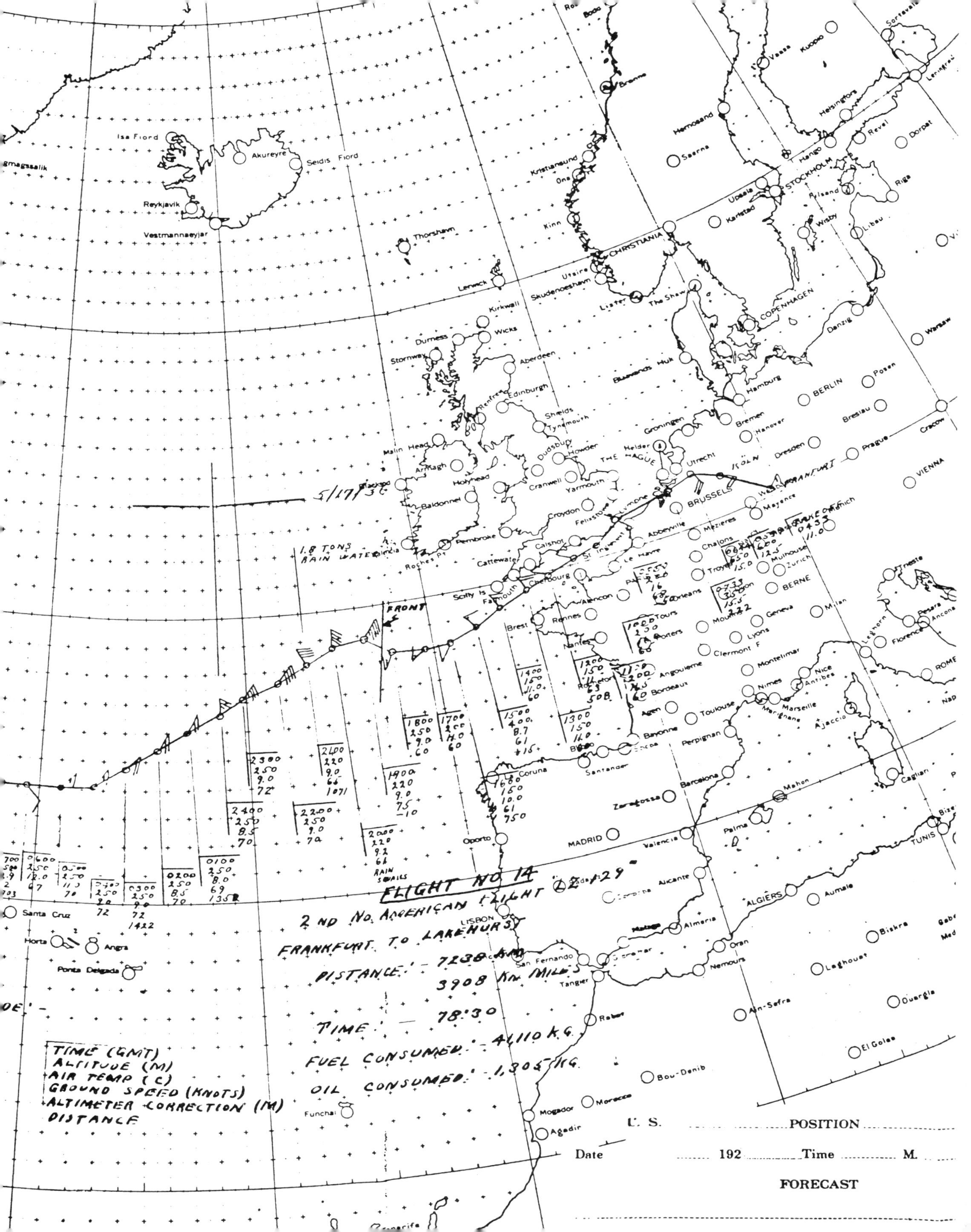

FLIGHT NO 14
2 ND NO. AMERICAN FLIGHT
FRANKFURT TO LAKEHURST
DISTANCE: 7238 KM
3908 KN. MILES
TIME: 78:30
FUEL CONSUMED: 41,110 KG.
OIL CONSUMED: 1,305 KG.
TIME (GMT)
ALTITUDE (M)
AIR TEMP (C)
GROUND SPEED (KNOTS)
ALTIMETER CORRECTION (M)
DISTANCE
1.8 TONS RAIN WATER
FRONT
RAIN SQUALLS
TAKE OFF
0433
11.0
5/17/36
Isa Fiord
Akureyri
Seidis Fiord
Reykjavik
Vestmannaeyjar
Thorshavn
Lerwick
Kirkwall
Wicks
Durness
Stornway
Aberdeen
Edinburgh
Malin Head
Armagh
Holyhead
Baldonnel
Pembroke
Scilly Is
Falmouth
Cherbourg
Croydon
Cranwell
Yarmouth
Brest
Rennes
Nantes
Tours
Orleans
Bordeaux
Angouleme
Agen
Bayonne
Toulouse
Perpignan
Santander
Coruna
Oporto
LISBON
MADRID
Zaragoza
Barcelona
Valencia
Alicante
Almeria
Gibraltar
San Fernando
Tangier
Rabat
Mogador
Agadir
Funchal
Santa Cruz
Horta
Angra
Ponta Delgada
Oran
Nemours
ALGIERS
Aumale
Biskra
Laghouat
Ain-Sefra
Ouargla
El Golea
Bou-Denib
Morocco
Palma
Mahon
TUNIS
Cagliari
Ajaccio
Marseille
Nimes
Nice
Antibes
Montelimar
Lyons
Clermont F
Geneva
BERNE
Zurich
Mulhouse
Milan
Trieste
Florence
Leghorn
Ancona
ROME
BRUSSELS
Abbeville
Mezieres
Chalons
Troyes
Mayence
Munich
FRANKFURT
KÖLN
Utrecht
THE HAGUE
Helder
Groningen
Bremen
Hanover
Hamburg
BERLIN
Posen
Breslau
Dresden
Prague
Cracow
VIENNA
Danzig
Warsaw
COPENHAGEN
The Shaw
Lister
Skudenoeshavn
Utsire
CHRISTIANIA
Kinn
Ona
Kristiansund
Bodo
Bronno
Hernosand
Saerna
Karlstad
Upsala
STOCKHOLM
Wisby
Libau
Riga
Reval
Dorpat
Helsingfors
Hango
Vaasa
Kuopio
Leningrad
Sortavala
U. S. POSITION
Date 192 Time M.
FORECAST

and the watch officer (Heinrich Bauer) elected to take the ship through. There was a relatively inexperienced man on the elevator control and he did not enter the turbulent area with the ship slightly down by the bow as he should have. The ship first pitched up ten degrees and then down ten degrees. The elevators were put hard down (eighteen degrees) and then hard up. Two men were required to hold the elevator wheel. Altitude increased 650 to 800 feet at a rate of 1100 to 1500 feet per minute. The engine mechanics in the after power cars, where they could see the lower fin, estimated that as the ship descended, and then responded to hard-up elevators, the fin came within 50 feet of the surface.

According to Dr. Eckener, who had entered the control car about this time, the ship should not have been taken into the front at this point, for there was insufficient visibility under the front and it could not be seen whether the opening extended all the way through under the front. His views concerning "assumptions" were again explained rather fully! The handling of the ship also was not too satisfactory and the elevator man was immediately replaced with a more experienced man who was standing by in the control car.

Head winds, some as high as force 9 and mostly from the southwest, were encountered until the ship was about five hours out of Lakehurst. At this time another front was encountered with winds shifting from southwest, force 7 (28–33 MPH) to northwest, force 7. Altitude when the ship encountered this front was 1000 feet. After passing through this front the winds varied from northwest, force 7, to west-southwest sometimes as high as force 9. Ground speed at times dropped as low as 35 MPH.

During the flight 7.3 tons of rain water were picked up by "brushing" against the rain squalls. By gaining this much ballast, the amount of hydrogen valved to keep the ship in equilibrium was greatly reduced. The engines operated perfectly throughout the entire flight and the only problem encountered was the weather, which truly was as bad as one would ever encounter.

Although the flight duration was 78 hours 30 minutes, the reserve fuel at landing was 13.05 tons, which was sufficient for 24.6 hours at normal cruising speed.

The return trip to Frankfurt on May 21–23 was almost as interesting as the flight to Lakehurst, for it was almost exactly the reverse, with good following winds practically all the way. Our flight time thus was only 48 hours 8 minutes. Where the westbound trip had been by the southerly route, going about as far south as the Azores, the return from Lakehurst was considerably farther north, the initial direction being northeast from Nova Scotia and Newfoundland, then north of Ireland, across the Irish Sea, then southeast over Howden and Yarmouth, across the English Channel to The Hague and on to Brussels and Frankfurt.

There were some published complaints about the ship's flying over England, although this route could not have been flown without permission from the British government. The complaints were about this being a "spy flight." But most of the flight over England was in fog and no one aboard the ship could see much of anything.

Winds the first day out of Lakehurst were from the northwest and sometimes reached force 6 (22–27 MPH). Winds gradually swung into the southwest reaching force 9 (41–48 MPH) the second day out. They gradually veered into the northwest, force 10 (49–56 MPH) as we approached the Irish coast. Here we encountered turbulent weather, the ship increasing altitude by 330 feet while ten degrees down by the bow and with two men holding the elevator wheel hard down, at eighteen degrees. Ship's speed was 75 MPH. The rate of climb was not as high as when entering the front on the flight from Frankfurt to Lakehurst.

The engines functioned perfectly and there were no shutdowns at all. Cabin ventilation was now considered satisfactory but better ventilation was still required in the smoking room. Altitude varied between 650 and 1500 feet until over the Irish Sea. At that time altitude was increased to 3000 feet and when the ship was over England the altitude was 1300 feet.

On this flight we had three U.S. Navy officers aboard as observers but only two bunks had been assigned to the four of us. We drew lots to see who would get first chance to use a bunk; I was unfortunate enough to get the last opportunity. Actually when my turn rolled around I chose not to do any sleeping for at that time we were encountering the winds of force 10 and turbulent weather. After reaching the hotel at Frankfurt I caught up on my sleep by sleeping all one day and the following night.

On this relatively short flight the *Hindenburg* consumed only 25.8 tons of fuel and still had as reserve on landing 29.5 tons or sufficient for 55.7 hours at cruising power. This was actually more than that used on the crossing.

At the end of this second North American flight, number 4 engine was removed from the ship in the Frankfurt shed and replaced by the spare engine. To remove the engine, three hours with four men working were required or the equivalent of twelve man-hours. Installation of the new engine required five hours with four men working or the equivalent of twenty man-hours.

Compared to conditions on the South American run, the weather over the North Atlantic could be violent, with gale force winds, marked temperature differences in air masses, and extreme turbulence with frontal passage. At the same time much more weather information was available to the ship's officers on this route than on the route to Rio.

Weather reports along the Atlantic seaboard of North America were received twice daily, at 0400 and 1600 GMT, from radio station NAA, which broadcast on several frequencies. These reports gave the weather at representative stations along the eastern half of the North American continent as well as ship reports as far east as Bermuda. If it was impossible to receive NAA when the *Hindenburg* was approaching the North American coast, the desired weather information was picked up by Angot Paris which relayed the NAA weather broadcasts. Ships' reports were broadcast by the Deutsche Seewarte in Hamburg at 0900, 1500, and 2055 GMT. The airship received these signals and the usual practice was to enter them on whichever map was of particular interest at the time—European, oceanic, or North American.

When weather information in a specific area was desired, the ship's radio station attempted to work any vessels known to be in that area. For this purpose a chart was kept aboard the airship which showed the position of all vessels crossing the Atlantic while the airship was under way. The observations of surface vessels were entered on the charts and also on the weather maps.

Under the term *Flugwetter* (aviation weather), general advisories were given at the request of the airship concerning the weather that might be encountered on a given course through certain areas. This was a German service provided by the Seewarte but the same type of forecast was also transmitted from the States on request. The *Grosswetterlage* (general weather synopsis) was also a Seewarte service and offered a description of the movements of air masses.

In general this system of receiving weather reports was satisfactory when the airship was in the vicinity of continents but the reports received over the ocean were not complete enough. Very little information was available concerning the weather northeast and east of Newfoundland. The maps drawn according to the *Flugwetter* and accompanying reports were very apt to show a simple pressure distribution while in reality the weather in this area was not nearly that simple, as flight experience showed.

The primary interest of airship personnel in the weather was with respect to the wind, both force and direction. Temperatures were relatively unimportant except as an aid in locating fronts and in that region over the North Atlantic where icing conditions might be encountered. Frontal conditions, always noted and located on the map when possible, were of particular interest in that they denoted a change in wind direction and intensity. The location of the front was observed and this aided the watch officers in their anticipation of bad weather.

The normal use of the weather maps as drawn aboard the airship was to lay out the ship's course in anticipation of the shortest possible flight time for the voyage in the light of the information available at the time. The course was not laid out to avoid frontal conditions for it was believed that no front was so continuous or so severe that there would not be a hole in it where the airship could pass through. It was often necessary to skirt the front for an hour or two, but sooner or later an opening would be found and the ship taken through. The hour or two lost here was much less than what would be lost if a longer course were taken to avoid the front.

Between May 25 and 29, 1936, the *Hindenburg* made her second flight to Rio. On this journey she carried as cargo a small airplane weighing 1.08 tons. The flight went straight through to Rio in 83 hours 13 minutes, and reserve fuel aboard on arrival was 12.15 tons, sufficient for twenty-three hours at normal cruising speed. The French had now given permission for the Germans to fly the *Hindenburg* over France and down the Rhone valley on the same route as used by the *Graf Zeppelin*. The same restrictions applied, however. The corridor down the Rhone valley was only twelve miles wide and since the turning radius of the ship was close to 5000 feet, there was not too much room for maneuvering around any thunderstorms in the valley. Still, the Rhone valley route was preferable to the longer route over the English Channel and around the northwest corner of France as taken by the *Hindenburg* on her first flight to South America.

"This flight," I wrote to Dr. Arnstein while en route to Rio, "is giving all of us an opportunity to catch up on our sleep while the weather is good, for none of us got much sleep on the flight to Lakehurst for there was too much to watch and we got even less on the way back to Frankfurt, for the ship was overcrowded and most of my sleeping was done in a chair or on the seat in the officers' mess. It's a great sport."

The return flight, May 30–June 3, 1936, was almost routine except that a thunderstorm was encountered north and east of Madeira. The existing conditions, and the route of the ship, are shown on the sheet of notes I made during this flight. After crossing the Mediterranean the course was set up the Rhone valley, then over Basel to Lake Constance. Engines were still functioning perfectly and the time between overhauls was set at 600 hours with consideration being given to raising it to 1000 hours. Duration of the flight was 94 hours 17 minutes, with reserve fuel at landing of 11.87 tons, sufficient for 22.6 hours at normal cruising. The ship carried forty-two passengers with fifty-three men in the crew.

The third flight over the North Atlantic to Lakehurst took place between June 19 and 22, 1936. Takeoff was at 20.52 GMT. Instead of going out over the Channel, the course was laid along the east coast of England with thunderstorms to the west of the ship's track. At Tynemouth the winds were out of the southeast at force 7 (28–33 MPH). The course was then laid almost due west across England and the northern tip of Ireland. In crossing England altitude was gradually reduced from 2460 to 985 feet. Thunderstorms were lying south of the ship's path as she flew north of Ireland.

From here the *Hindenburg* steered directly for Newfoundland. Altitude over the North Atlantic varied from as low as 600 feet to as much as 4000 feet in seeking favorable winds. Although fog and rain were at times encountered, winds for the most part were fairly light until 2200 on June 20 when south southeast winds of force 7 were picked up. From 2300 to 0100 the winds shifted from south southeast, force 7, to north northwest, force 6 (22–27 MPH). The ship was taken from 2000 feet to 4000 feet and back to 920 feet. Head winds and fog were then encountered as the *Hindenburg* approached Newfoundland and the ship's location was determined by radio compass, a correction of eighty-seven miles having to be made in the ship's position. Crossing Newfoundland north of Cape Race, the ship's altitude was 1200 feet.

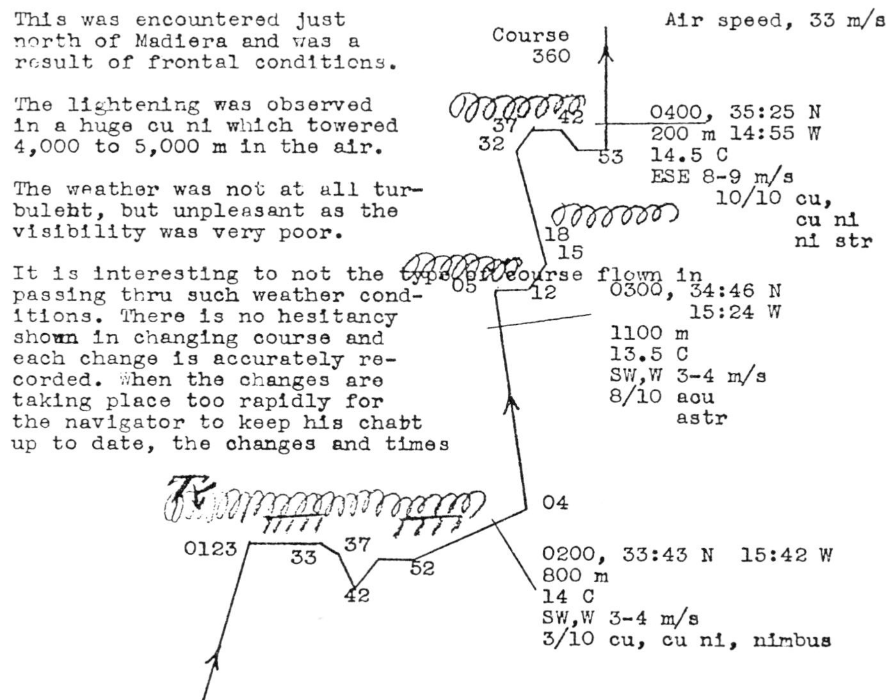

SECOND SOUTH AMERICAN FLIGHT OF LZ 129
Rio de Janeiro to Frankfurt,

Thunderstorm:

This was encountered just north of Madiera and was a result of frontal conditions.

The lightening was observed in a huge cu ni which towered 4,000 to 5,000 m in the air.

The weather was not at all turbulent, but unpleasant as the visibility was very poor.

It is interesting to not the type of course flown in passing thru such weather conditions. There is no hesitancy shown in changing course and each change is accurately recorded. When the changes are taking place too rapidly for the navigator to keep his chart up to date, the changes and times are merely recorded and the chart brought up to date when the weather has quieted down a bit. Normally there is sufficient time and the navigator will never be more than two or three changes behind in his plot of the course. Drift readings are constantly being taken under such conditions, which with such frequent changes in course, makes it possible for the determination of the wind, direction and velocity, at all times.

A sheet of notes made by Harold G. Dick on the *Hindenburg*'s second South American flight, return leg, May 30–June 3, 1936, indicating courses followed through a thunderstorm north of Madeira. (From the collection of Harold G. Dick)

While the ship was flying in the fog her sonic altimeter was in constant use. This was a much improved instrument over that used by the *Graf Zeppelin*. It was operated by compressed air with a very audible "beep" when the signal was transmitted and a somewhat less audible "beep" when the echo returned to the ship. At an altitude of 1200 feet there would be about two seconds between the two beeps at sea level and this would be reduced by the height of the terrain over which the ship was flying. While we were crossing Newfoundland the beeps were slightly less than two seconds apart as one would expect, for most of that region is not very high above sea level. The only hill in that area must have been right on the ship's course, for suddenly the beeps came closer and closer together. The ship's altitude was increased as quickly as possible, and as the *Hindenburg* passed over whatever was below, the two beeps combined into one continuous longer beep. How near the ship came to the surface nobody will ever know because of the fog and poor visibility— but it had to be very close.

Fog and head winds, sometimes as high as force 9 (41–48 MPH) were encountered as far as Halifax, but the winds dropped off thereafter and then built up to following winds all the way to Lakehurst.

At 0650 GMT on June 22, and about two hours out of Lakehurst, the ship flew over the giant new Cunard liner *Queen Mary*, which had sailed on her maiden voyage from Southampton on May 27, 1936. This was shortly after midnight local time and the *Queen Mary* had all her lights ablaze as the *Hindenburg* flew over her.

At Lakehurst I disembarked, and remained in the United States until my return to Germany on the *Hindenburg*'s last North American flight of the season in October. I was then on loan from Goodyear-Zeppelin to the German operating company, the Deutsche Zeppelin Reederei, to assist in their activities at Lakehurst.

Visitors were not allowed on the field during the landing maneuver and some Americans of German descent made quite an issue of it. There was an organization, the German-American Bund, headquartered in the Yorkville section of New York City, that was patterned after the Nazi Party in Germany even to the point of wearing black riding breeches like the SS Black Shirts or storm troopers, except that the Americans wore white shirts. They appeared at Lakehurst in quite a large group and insisted that they be allowed on the field and close to the ship during the landing. Since this was against the rules, I refused to let them out on the field.

They assumed that I was one of the Germans and reported me to Nazi Party headquarters in Berlin. This complaint was then routed from Berlin to the Zeppelin works in Friedrichshafen, and I heard quite a bit about this when I returned in October. But the complaint was not taken very seriously. In fact, it was considered something of a joke that I had stepped on the American Nazis' toes and it probably raised my stock a bit that I had not given in to their demands.

I returned to Germany on the *Hindenburg*'s last flight of the season from Lakehurst to Frankfurt, between October 10 and 12, 1936. The time en route was 52 hours 17 minutes. This was my tenth crossing of the Atlantic in the big ship, six crossings having been over the North Atlantic and four over the South Atlantic. I was interested in the changes made in the ship and its operation during the short period of my absence since the end of June.

Additional personnel were being carried in order to train as many men as possible. On this flight the crew consisted of forty-seven men and in addition there were thirteen stewards. Rudolf Sauter, as chief engineer, was responsible for maintenance of the ship and did not stand a watch. He had three assistant engineers, each of whom stood a four-hour watch.

Automatic steering (for rudder control only) was used extensively but only in calm weather. In rough weather manual control was used. The rudder man reported for duty as scheduled but if the weather were calm the automatic steering would be used and he would take over the elevator control, relieving the elevator man who could then be assigned other duties. It was

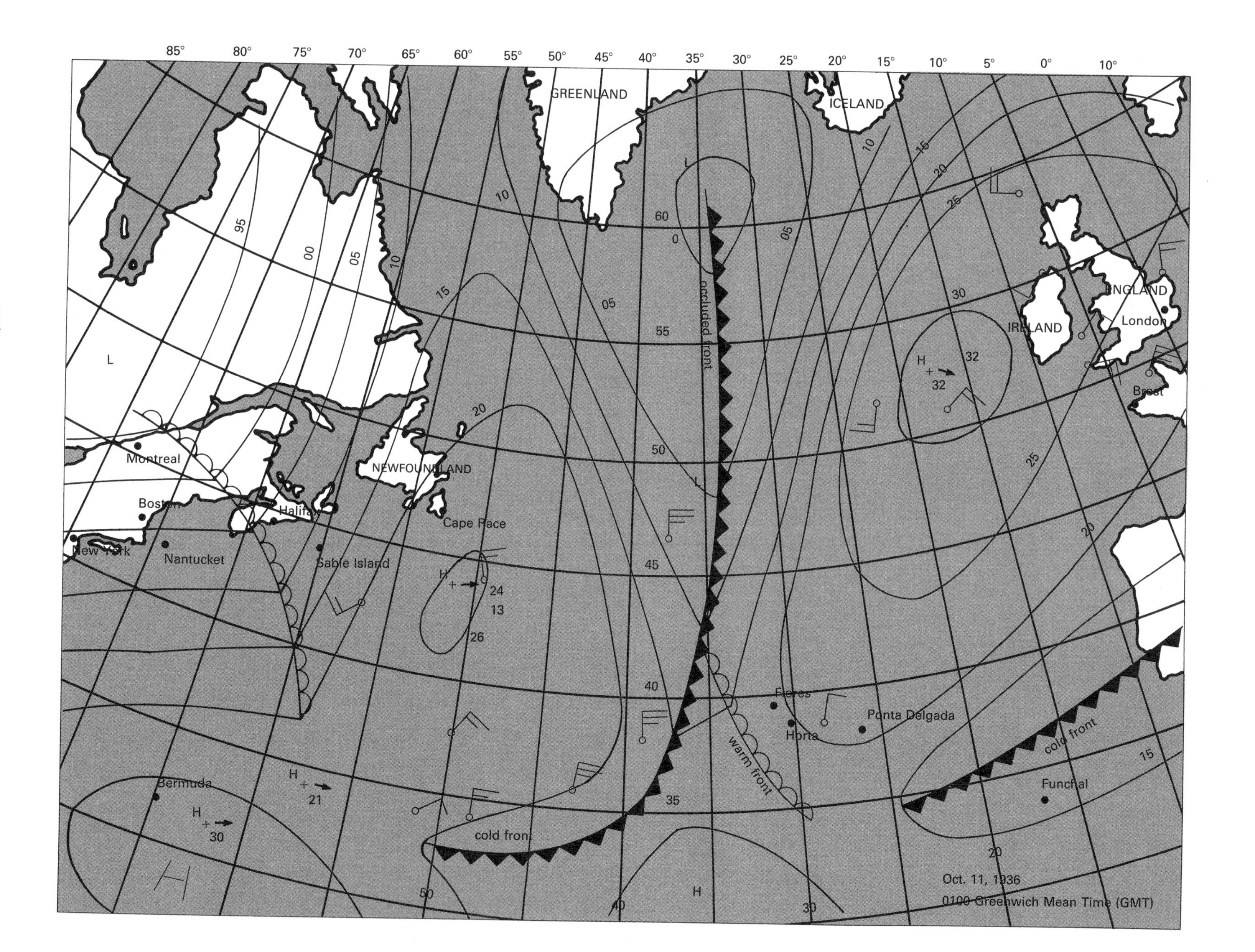

85°
80°
75°
70°
65°
60°
55°
50°
45°
40°
35°
30°
25°
20°
15°
10°
5°
0°
10°
GREENLAND
ICELAND
ENGLAND
London
IRELAND
Brest
NEWFOUNDLAND
Cape Race
Montreal
Boston
Halifax
New York
Nantucket
Sable Island
Flores
Horta
Ponta Delgada
Funchal
Bermuda
occluded front
warm front
cold front
cold front
Oct. 11, 1936
0100 Greenwich Mean Time (GMT)

ON PAGE OPPOSITE: Weather map showing conditions encountered on the *Hindenburg*'s flight from Lakehurst, New Jersey, to Frankfurt, on the second day of the flight, October 11, 1936. (From the collection of Harold G. Dick)

said that in calm weather the automatic control held the ship on course better than could be done manually. Because of this, consideration was being given to possible reduction in personnel.

All instruments were functioning satisfactorily except that the airspeed meter read too high. When it read 76 MPH, the navigator was logging 74.9 MPH.

The automatic valves for cells 2 and 3, which had been a problem on the return from the first flight to Rio, were still not operating properly, vibrating when they started to open. The procedure, started in the spring, of valving gas manually from these cells when pressure height was being approached was still being used.

Strain gauges had been mounted in the lower fin at frame 47.5 and also in the lower fin at frame 38, on the top side of the port horizontal fin at frame 47.5, and on the bottom side of the port horizontal fin at frame 33.5. A total of eighteen strain gauges had been installed and were being monitored in flight by three members of the DVL and the Luftschiffbau Zeppelin stress analysis department. They reported that the highest readings so far obtained were of the order of 5600 pounds per square inch.

The second morning out of Lakehurst we overtook a cold front of considerable intensity. The wind shifted from northwest 25 to 27 MPH to south-southeast 40 to 45 MPH, with a temperature increase of 5 degrees F. The wind shift was very abrupt, taking place in a little less than a minute. It seemed that the two air masses were not coming together at all, but simply moving past each other. When I discussed this with Dr. Eckener, he stated that it was like a curtain and about as sharp a line of demarcation as he had ever seen. The turbulence was not severe and did not come anywhere near the turbulence we had experienced on the second flight from Frankfurt to Lakehurst.

The *Hindenburg* had by now accumulated about 2000 hours' flying time and it could be said the ship was functioning very well. There had been a minimum of maintenance problems, wire breakage, or other difficulties. The ship was scheduled to make three more round trips to Rio, which would make a total of seven round trips to Rio and ten to Lakehurst. By the end of this first season the ship would have flown a little less than 3000 hours.

On the three remaining flights to Rio, Captain Max Pruss was to relieve Captain Lehmann as commander of the *Hindenburg*. I remember that when Captain Pruss was in command of the ship, every landing was perfectly executed.

Chapter 13 *Plans to Expand the Transatlantic Airship Service*

I was surprised by the changes in the atmosphere in Friedrichshafen during my absence of only a few months. One obvious and understandable alteration was the confident, even arrogant, attitude of everyone connected with both the Reederei and the Luftschiffbau Zeppelin. In the spring of 1936, with the *Hindenburg* just starting her trials, they were sure they had a good ship equal to the challenge of regular service across the North Atlantic. At the same time, no aircraft had ever attempted such an operation, and there were doubts and questions that could be answered only by experience. At the end of the operating season, October 1936, they were now enjoying a dazzling success in maintaining an accident-free service, with enormous amounts of favorable publicity, not only in the newspapers of the world but in the excited testimony of delighted passengers.

Governments and big business were considering, even accepting, the proposition that the era of transatlantic air travel had arrived, and that the ideal vehicle was the big rigid airship. Confident that in the *Hindenburg* they had produced an outstanding type of airship, the Germans planned to institute an accelerated building program that would add at least two more similar passenger ships to the Reederei's fleet. Already a slightly modified sister to the LZ 129, the LZ 130, was taking shape in the building hangar in Friedrichshafen.

When I wrote to Mr. Litchfield on November 9, 1936, concerning German plans for expansion, I advised him that the LZ 130 was expected to be completed in September 1937. As soon as possible she would be transferred out to the hangar at Löwenthal to permit construction to commence on the LZ 131. With an added 59-foot bay, and lift to carry one hundred passengers, this ship, with a length of 862 feet, would be too long to fit in the 820-foot Friedrichshafen construction shed. The answer was to lengthen the shed to 917 feet. It was proposed to start work on one end of the hangar, while simultaneously the erection of the LZ 131 would commence at the other end of the hangar with the frames just forward of the fins, and erection would progress towards the bow. By the time the hangar reconstruction was complete, the forward structure of the LZ 131 would be finished and the main frames supporting the fins would be ready for erection. LZ 131 would then be ready to fly in July 1938. At the same time, a sister ship, the LZ 132, would be under construction in the Löwenthal hangar, which already measured 885 feet in length. A ring building shed would have to be erected at Löwenthal, while the ring building shed at Friedrichshafen was to be lengthened from 492 feet to 885 feet.

A corresponding development would have to take place at the overseas operating base at Frankfurt, where a single large hangar 902 feet long, 170 feet wide, and 167 feet high had been completed in May 1936 for the *Hindenburg*. A second hangar of the same dimensions was to be built and finished by May 1938 for the LZ 130,[1] at a 30-degree angle to the first one. A third hangar was planned, a revolving "docking shed" located at the point of intersection of the center lines of the two fixed hangars. Turned parallel with the prevailing wind, the docking shed would permit airships to enter or depart regardless of wind direction. Capable of lining up with either fixed shed, the docking hangar could convey the airship from one shed to the other regardless of weather conditions.

Many personnel changes were to accompany the expansion of the Reederei fleet. Captain Lehmann, as director, was to stay ashore administering personnel and operations. Hans von Schiller, formerly commanding the *Graf Zeppelin,* was to be captain of the LZ 130, while Sammt, Milce, and Eichler would be watch officers. Pruss would continue in command of the *Hindenburg* with Heinrich Bauer, Ziegler, and Zabel. Milce, Eichler, Ziegler, and Zabel were all new men, ship's officers from the HAPAG (the Hamburg Amerika shipping line) who came to the Luftschiffbau Zeppelin or the Reederei to start flying in 1935. Though they had all flown about 5000 hours by the end of 1936 and knew their jobs thoroughly, it was generally considered that they were young in experience, and the burden of responsibility would fall on the shoulders of the oldtimers— Pruss, Von Schiller, Sammt.

Meanwhile the old *Graf* was to be retired from passenger service, according to the plan, continuing to fly as a school ship. Anton Wittemann was to be the commanding officer, with Hans Ladwig as watch officer. Making short training flights, it was thought she would require only one watch of old hands.

Dr. Eckener in particular had always expected that an American company would cooperate in passenger airship operations across the Atlantic, the Germans flying two ships and the Americans two. In the immediate aftermath of the *Graf Zeppelin*'s world flight, the International Zeppelin Transport Company had been formed on October 18, 1929, and had retained as its vice-president and general manager Jerome C. Hunsaker, who as a former naval constructor had played a prominent role in the early development of rigid airships in the U.S. Navy. It was Hunsaker's job to study the problems and economics of a transatlantic airship service. No action had been taken on his recommendations because of the Depression. In 1936, as a result of the *Hindenburg*'s success and the renewed interest in carrying passengers by airship, the International Zeppelin Transport Company was revived and renamed the American Zeppelin Transport Incorporated, with Edward P. Farley, chairman of the American-Hawaiian Steamship Company, as president and F. Willy von Meister, in America as Dr. Eckener's personal representative, as vice-president.[2] Several prominent industrialists were on the board of directors, including President Litchfield of Goodyear Tire and Rubber, Edward A. Deeds, chairman of National Cash Register Company, and Charles F. Kettering, vice-president of General Motors. Dr. Eckener was the only officer who was a foreigner. He was serving on the board by invitation so that his wide experience would be available to the American counterpart of the Deutsche Zeppelin Reederei.

Prior to 1936 it had always been assumed that Goodyear-Zeppelin would design and build the two American ships specifically for helium operation. But by the end of the 1936 season, the Germans, filled with justifiable pride in the accomplishments of the *Hindenburg*—plus in certain people some of the nationalistic chauvinism so characteristic of the Nazi period—were arguing that the American ship should follow the German prototype. Associated with this point of view was a condescending attitude towards Goodyear-Zeppelin, with the feeling that they had accomplished very little since the completion of the *Macon* in 1933 and had received far more technical information from the Luftschiffbau Zeppelin than they had provided in return. I ran into this attitude as early as February 1936, when I had to write to Mr. Litchfield that Dr. Dürr had refused to provide me with some technical information. "Since Goodyear-Zeppelin is a paper organization according to his understanding, [Dürr] does not feel free to give out information of a highly technical nature until the situation is cleared up. This also involves the question of patents. He has therefore asked that he be given a little more information concerning the status of Goodyear-Zeppelin and particularly how such things as patents and the interchange of information is to be handled until Goodyear-Zeppelin is again an operating concern." I added, "my personal reaction is that there is some information that Luftschiffbau Zeppelin is not so keen on interchanging with Goodyear-Zeppelin and the present status of GZ offers a fine excuse."

For the American firm to build to the *Hindenburg* design would have created many problems. The German plans used the metric system throughout and all design drawings would have had to be converted to feet and inches. I checked many of their drawings and found that dimensionally this was not an insurmountable problem but many fittings would have to be proof-tested before they could go into production. Dies were more of a problem. Although the die drawings were quite complete, practically all the dies had been changed to make them operate properly without incorporation of these changes into the drawings. Dr. Dürr's recommendation as to the cheapest and most satisfactory arrangement was that Luftschiffbau Zeppelin should make the dies and sell them to Goodyear-Zeppelin.

Another consideration to which the American company should have paid heed was Dr. Eckener's stated conviction that it would take two years to train an American crew to competently handle the American commercial craft. At the same time, I believed that it would take at least two years for Goodyear-Zeppelin to get their first ship into the air, simply because of the need to assemble new equipment, tools, jigs, dies, that would be needed before we could even start construction. Some of my German friends advised that we should buy our first ship from the Luftschiffbau Zeppelin and build the second at leisure. On November 9, 1936, I recommended to President

Litchfield that "chartering a ship seems to me to be the only way to get started and for definitely training a crew." Some of the Germans felt that the Reederei needed all the airships it could get and therefore could not afford to sell any. A contrary idea was that airships would be built faster than men could be trained to fly them and faster, too, than terminal facilities could be erected to house them.

A further change involved the extension of German Air Ministry control over the Luftschiffbau Zeppelin, accompanying the greater involvement of the company in the rearmament program. This had taken place during the summer and fall of 1936. No drawings or data dealing with airships or any aspect of the aircraft industry could be released without approval of the Air Ministry. Further, restrictions were placed on the number of outsiders like myself in Friedrichshafen. Following my return to Friedrichshafen in October 1936, Dr. Eckener had to take up the matter of my presence at the Luftschiffbau Zeppelin with Staatssekretär Erhard Milch, Goering's right-hand man at the Air Ministry in Berlin. Three weeks passed before the Air Ministry advised Dr. Eckener that one person (in this case myself) could remain for six months at the plant, when another person might take my place if I went back to the States. As far as Dr. Eckener was concerned, the relationship was to continue as before, while as far as I was personally concerned, I had been so long in Friedrichshafen that I was "like an old piece of furniture" to the people at the Air Ministry. There was no question about my operation except that there could be no official release of information. Still I was free to pick up all the information I could—unofficially.

In the middle of December I spent a few days in Frankfurt, where the *Hindenburg* was laid up for the winter. I timed my visit so as to see gas cells 3, 11, and 16 being removed from the ship for inspection. A lot of work was being done on the ship, mostly on the order of inspection and cleaning; very few changes were being made. All four diesel engines were removed and sent for modernization to the Daimler works at Untertürkheim. Also, because of the demand for passenger space on the North Atlantic run, ten more cabins were to be added in bay 11 along the keel abaft the regular passenger quarters. The new cabins would accommodate twenty-two more people (nine were two-berth cabins, while the tenth cabin accommodated four people and was intended for occupancy by a family with children). The new cabins had an outside view via the long row of 48-inch-wide windows along the bottom of the ship.

I also saw a new propeller of a novel material—thin laminated wood cemented together under great pressure so that the wood was really impregnated and became as hard as dural and could not be cut with a knife. The new propellers were also 19.7 feet in diameter, with three instead of four blades; because the blades were machined and inserted in a metal hub, they were ground adjustable. Efficiency was supposed to be 73 percent instead of 68 percent, as the new blades were immune to flutter and the increased stiffness permitted the use of thinner and more efficient sections.

Chapter 14 *Hook-On Experiments in the Spring of 1937*

he *Hindenburg*'s first flight of the 1937 season was scheduled for March 11, 1937. At last the Luftschiffbau Zeppelin had made use of the U.S. Navy drawings of the American trapeze and hook-on equipment, which had been procured the year before with the cooperation of Goodyear-Zeppelin. On this first flight of the year, the trapeze was installed and was to be tested. General Ernst Udet—with sixty-two victories the highest-scoring German ace to survive World War I, and now Director General of Equipment for the Luftwaffe—was to pilot the hook-on aircraft, a light Focke-Wulf Stieglitz training biplane with a 150-HP engine and loaded weight of 1433 pounds. Rightly esteemed between the wars as a brilliant stunt and airshow pilot, Udet had ability and skill beyond question. I went to Frankfurt on March 10 especially to watch the hook-on test. Captain Pruss seemed glad to have me along. Even though I myself was not too familiar with U.S. Navy hook-on procedure, I was at least in a position to relay any questions back to the States.

I have been told that historians find it impossible to obtain any information on these hook-on trials, so I will give my observations in some detail as they were passed on in my letters to Dr. Arnstein:

> In total, 5 hook-on landings were made [during the March 11 flight]. From the difficulty which Udet was experiencing in making the hook-on, it was apparent that there was something basically wrong with the set-up. Before he succeeded in making the first landing he was in contact with the trapeze at least 3 or 4 times. Each time the ship experienced a slight shock and the plane bounced off. The following 4 hook-ons cannot be said to have been much better.
>
> From the discussions that took place after the ship was again in the hangar and that evening, I would draw the following conclusions:
>
> The trapeze seemed to be located where there is some turbulence at ring 140, for the plane was very steady in the air about 1 meter below the trapeze and also a little further aft. Further aft it was possible for Udet to get much closer to the hull than was actually necessary and still without experiencing this turbulence.
>
> Just forward of the trapeze there is a comparatively loose section of outer cover, that over the large freight compartment, and it is possible that this may have caused some of the turbulence. It may also have been caused by the control car.
>
> The hook was so rigged on the plane that the entire hook and guard was not visible. The pilot could see the forward part of the guard and the hook but the center was blind. . . .

As a result of these tests, the guard and the hook were somewhat damaged.

The *Hindenburg* started on her first ocean crossing of the year to Rio under Captain Pruss's command on March 16, 1937. The new cabins along the keel in bay 11 were not complete, but it was expected that they could be finished in April, before the season's first North American flight, due to depart from Frankfurt on May 3, 1937. In between, another test flight was scheduled for April 27 during which hook-on landings were to be made. Since the trapeze and hook-on plane had both been altered, I wished to take part in the flight and again Captain Pruss immediately gave his consent.

Udet was again scheduled to fly the plane, but if the first tests were successful one of the Focke-Wulf pilots was to replace Udet. The tests didn't get that far.

The hook on the plane was not greatly altered. The struts supporting the hook and guard were damaged during the initial tests last month so in replacing them the height of the hook above the wing was somewhat increased, but not a great deal, approximately .25 m.

The trapeze was completely rebuilt, but was kept in its old location, that is at ring 140. The overall length of the trapeze is now a little less than 3 m and the trapeze has been divided into two parts, the lower section being free to swing forward when in contact with the plane, but at the same time this lower section is held from dropping back by shock absorbers, one on each side of the trapeze. Instead of the brakes that were formerly used where the trapeze is hinged on the ship there are now hydraulic shock absorbers, the brakes having been completely removed.

The tests were made about noon in weather that was not particularly good, overcast with occasional light rain, and it was not as quiet as when the first tests were made last month. Otherwise conditions were not altered.

Udet managed to make four hook-ons but it could be seen that he was having a lot of difficulty although it appeared that the longer trapeze was a help. After he had made the fourth hook-on a light gust which we did not feel on the airship apparently struck the plane carrying it across the trapeze with considerable yaw and a lot of roll. Udet pulled the release, but the propeller tips struck the trapeze, were damaged, and he had to make a forced landing. The plane was furnished with a new prop and flown back to the airport.

In the afternoon when the weather was a bit quieter the tests were again repeated with no better results. Two landings were made on the trapeze and then the same thing happened again that had happened in the forenoon. This time the propeller was quite seriously damaged and another forced landing was made. A new prop was again furnished and the plane flown back to the airport, but the tests were finished for the day.

I had breakfast with Udet at the hotel both before these tests and again on the following day and he is thoroughly convinced that the difficulty lies in the turbulence that is encountered at this position on the ship. He claimed that it is so difficult to execute the maneuver that each landing can be classified as a work of art.

There also seem to be some difficulties introduced by the men in the ship keeping the hatch near the trapeze open. That may make matters worse but it is not the solution for on the first set of tests last month the hatch was at times closed. A window in the outer cover at this position would unquestionably be better.

Since in both instances when the propellers were damaged the plane moved

General Udet makes a hook-on landing to the trapeze on the *Hindenburg* in the spring of 1937. Although a number of hook-ons were made, the tests were unsuccessful. According to Udet, each landing had to be a "work of art."

> across the trapeze from starboard to port and since the trapeze is on the starboard side of the ship, between longitudinals number 0 and number 1, it may be that the off-center position of the trapeze adds to the difficulties. . . .
>
> The light plane that is now being used is also open to question as the weight is considerably less than that of the planes we used and the wing loading is also quite low. It is possible that at a later date a heavier and more powerful plane will be used.
>
> Just what will be done now I do not know. I have forwarded some of Udet's questions to Comdr. Rosendahl and the reply may have considerable effect on the method of attacking the problem.
>
> At the present time the idea of picking up mail, etc., is out of the question and some radical change must be made in the setup before the hook-on landing can possibly be carried out with any degree of reliability.

In short, despite Udet's great skill and experience, the hook-on experiment was a failure and this explains why historians find nothing concerning it in German sources. What might have been done in the future is a question, as all prospects of further hook-on tests ended ten days later with the burning of the *Hindenburg* at Lakehurst.

After each series of tests our group discussions ran into the evening and usually ended up over a glass of beer. But I enjoyed even more discussing the problems with General Udet at breakfast before and after the tests at the Frankfurter Hof, where we were both staying. Here there were just two of us trying to figure out why he had so much trouble when U.S. Navy pilots were so successful in making the hook-ons and releases on the *Akron* and *Macon*.

At no time in any of these discussions was there any discussion of Nazism. Nazism was not even mentioned. General Udet's response to the "Heil Hitler!" greeting seemed a little less than halfhearted. I wondered what might have happened to Udet, for my feeling was that he did not get along too well with the Nazis. It was a real shock when I learned of his end.[1]

Forty years later I met Admiral Harold B. Miller, who as a lieutenant flew with the hook-on heavier-than-air units of the *Akron* and *Macon*, and was senior aviator of the *Macon* in her last year. From him I learned for the first time of plans to make hook-on flights to the *Hindenburg* on the western side of the Atlantic. While the tests were in progress in Germany, Miller was designated as the pilot of the hook-on plane that was to intercept the *Hindenburg* off the New England coast on her first flight to Lakehurst in 1937. Had the tests been successful he would have met the airship, carrying two customs officials who would have boarded the *Hindenburg* so that the passengers could clear customs before landing. Meanwhile Admiral Miller would have returned with the mail from the *Hindenburg*.[2] It's a small world!

Chapter 15 *The Disaster at Lakehurst*

hy was I not aboard the *Hindenburg* on May 3, 1937, when she departed from Frankfurt under Captain Max Pruss's command for her first North American flight of the 1937 season—the flight that ended so tragically and spectacularly three days later in the totally unexpected fire at Lakehurst? I certainly would have wished to observe her performance and compare it with that of the previous year, and would have been readily accepted if I had asked to participate. However, on April 13, 1937, I had received from President Litchfield a letter:

> My dear Hal,
>
> I haven't written to you for quite a while but I have been very much interested in receiving your letters concerning the progress of construction, and your weekend trips into the mountains. You said that you plan to scale a new mountain every week, and as there are quite a number you haven't scaled, it wouldn't seem necessary for you to return here very soon.
>
> As a matter of fact, airships are making very slow progress here, and I think you could best employ your time by remaining in Friedrichshafen until conditions change here. In your present assignment you are probably acquiring more experience which will benefit you in the future than you could anywhere else at present.
>
> I am glad you are getting along well, and I hope to see you some time in May. I am planning to be in England between May 9 and 18, spending most of the time in England and a few days in Holland. I doubt if I will get to Friedrichshafen, so I would like you to come to either Amsterdam or London to see me. . . .

I of course abandoned my plans to make the May 3 flight, and on receiving a later telegram, arranged to meet President Litchfield in London between May 11 and 15. Thus I missed participating in the fatal flight. My guardian angel was working overtime and I have been very grateful ever since.

I have often been asked how my folks found out that I was not on that last, disastrous flight of the *Hindenburg*. With communications being as slow as they were at that time, my folks never knew whether I was in Friedrichshafen, Frankfurt, or Berlin, or—if the *Hindenburg* was flying—on my way to South America or to Lakehurst.

My brother Ernie, who was located in Mansfield, Ohio, at that time, called Ed Thomas, who was president of Goodyear, and that started a series of telephone calls to Lakehurst. About two hours later information was obtained from one of the crew members who had been aboard the *Hindenburg* that I had not been aboard the ship with them. This then was relayed back to my mother and dad who were in New England. They obviously had had a few hours of waiting that was not very pleasant.

About four o'clock in the morning I received the news of the *Hindenburg* casualty, which seemed like a bad dream until morning when there was no question about its reality. I immediately sent a cable home to my family in New England, stating simply "Still here in Friedrichshafen." At the time I thought that was all the information that was necessary, but I've since taken a lot of kidding about the "long" telegram that I sent.

Friedrichshafen was a town in mourning. Twenty-two *Hindenburg* crew members were dead, including Ernst Lehmann, prewar DELAG captain, war hero, Dr. Eckener's right-hand man in the 1920s, and director of the Reederei. In the small town on the Bodensee all of the men were well known, and left many grieving relatives. On May 21 I was in Cuxhaven, along with Knut Eckener and a few others from the Luftschiffbau Zeppelin, for the state funeral for the dead crew members, after which there were the burials in Friedrichshafen. Dr. Eckener and Dr. Dürr had left to testify at the inquiry at Lakehurst. Meanwhile there was great concern for the survivors, many gravely injured, in hospitals in the United States. For a time it seemed that Max Pruss, the *Hindenburg*'s commander, might not pull through. He did live, though he was in and out of hospitals for skin grafting and other surgery for many months, even after his return to Frankfurt.

From the conflicting information that came in, we found it impossible to reach a conclusion as to the cause of the fire. "Each bit of information we receive seems to eliminate another of the various possibilities," I wrote to Mr. Litchfield on May 24. "We now know that gas had not been valved for several minutes preceding the casualty, that the landing was normal in every respect, that there was no major structural or mechanical failure. The possibility of an engine backfire igniting the gas we have always discounted because of the low temperature of the diesel exhaust. It is becoming increasingly difficult to even advance a theory as to what may have ignited the hydrogen."

As I look back almost fifty years at the loss of the *Hindenburg*, it becomes evident that there was not—and probably never will be—proof that the ship was destroyed by static electricity, by some form of structural failure, or by sabotage. The views of reputable airshipmen are at variance, with absolutely no uniformity of thought.

The sabotage theory is the most sensational and the easiest to understand. As Dr. Eckener said to me, "anyone can understand sabotage—some crazy person does some crazy thing. On the other hand if the loss were caused by some natural phenomenon, considerable testing and research might be necessary to determine what actually happened." Believing, apparently, that the loss was caused by a natural phenomenon, that is, by static electricity, the Germans built a full-scale mock-up of that portion of the hull where the fire first broke out, the area just forward of the upper fin, and installed it in an electrical laboratory in Germany and simulated all the conditions that existed at Lakehurst at the time of the casualty.

The conclusion arrived at by the blue ribbon panel was that the tests showed there had been a discharge of electricity from the moist outer cover to the grounded structure, thereby igniting some free hydrogen in that area. The contributory conditions were that the outer cover was in effect insulated from the hull structure by dry predoped lacing cord, the hull structure was grounded through the handling lines that had been dropped and were already hooked up to the cables of the hauling down winches, the free hydrogen could have resulted from a leaking gas cell or trapped gas that was not properly vented during the weigh-off procedure, and there was a high potential gradient in the atmosphere that put a high static charge on the moist outer cover. This version of the cause of the casualty was officially accepted and accordingly published but it was not accepted by many of the airship personnel.

Rudolf Sauter, Chief Engineer aboard the *Hindenburg,* did not agree with the official findings. He stated that there was extensive St. Elmo's fire during the tests, which he said did not exist at the time of the casualty.[1]

Max Pruss, who was in command of the *Hindenburg,* maintained that the loss had to be from sabotage.

Admiral Charles E. Rosendahl of the U.S. Navy believed the cause to be sabotage and up to the time of his death was still hoping to find a positive indication of sabotage, such as batteries for an infernal device.

There were really four conditions that could have provided the free hydrogen necessary for the *Hindenburg* fire. A small leak, tear, or chafed area could have released a small amount of gas without noticeable loss of lift in the aft portion of the ship. A valve could have stuck open. The valved gas from the weigh-off might not have been properly vented. The fourth condition is a torn cell due to a broken wire.

The first is, I believe, relatively unimportant. The ship had a lot of hours on it and the winter inspection had showed some chafing so there could have been some of that type of leakage in the past and no resulting problems.

A valve sticking open would have been noticed on the indicators in the control car. Although there was one case (on leaving Rio on the first South American flight) when an overpressure valve stuck open, I know of no instance when a *maneuvering* valve stuck open. Because it hadn't happened, it doesn't mean it couldn't happen, but I would discount this.

Venting of the gas through the hoods was very good. The chimney effect, even at a very low speed or no speed, vented the gas very well. I would also discount this.

From what I have heard there was a definite ripple in the outer cover during the landing maneuver of the *Hindenburg.* Considering its location it would have to come from a torn cell. Wire breakage would be experienced in the gas cell wiring, not the shear wires, and would have little effect on the ship's strength, but could tear a gas cell—as is indicated in this case. For a torn cell to ripple the outer cover there would have to be a lot of released gas and with a lot of gas there would have to have been some with just the right mixture of gas and air to be ignited. To me, everything points in this direction.

Some persons attempted to blame Captain Pruss for the accident in that he allegedly persisted in his approach with a ship that was tail heavy and not in trim. The most stable condition for an airship was in trim but just slightly heavy, meaning that it would fly slightly, one or two degrees up by the nose. Under these conditions the ship would almost fly itself while making the landing maneuver. As speed was reduced the angle would be increased. Pruss used this method and made beautiful landings. Coming in that way the ship would appear slightly tail heavy and this could explain why some people at Lakehurst thought the ship was tail heavy and in trouble.

Overriding the technical approach in seeking the cause of the casualty is the fact that the *Hindenburg* was operating under the Nazi flag of Germany, and many of the operating personnel were strong members of the Nazi party. It was generally believed (Rosendahl as an example) that the Nazi hierarchy did not accept the sabotage theory because of the effect on their prestige. A number of the *Hindenburg* personnel were strong, ardent, and high-ranking Nazis but even so they openly stated their belief in sabotage.

Furthermore all the *Hindenburg* operating personnel had tremendous faith in their ship. They believed it to be a good and sound airship. It had already made ten round trips over the North Atlantic and eight round trips over the South Atlantic to Rio with a minimum of problems. It was from any point of view a sound ship. There was also an excellent esprit de corps aboard the ship. The flight personnel were hand picked and their ability to get along together was an important requirement. If sabotage were the cause, someone other than a member of the crew would probably have been responsible.

There are two items not in common knowledge. When the outer cover of the LZ 130 was to be applied, the lacing cord was prestretched and run through dope as before, but the dope for the LZ 130 contained graphite to make it conductive. This would hardly have been necessary if the static discharge theory were merely a cover-up. The use of graphited dope was not publicized and I doubt if its use was widely known at the Luftschiffbau Zeppelin.

Additionally, when the LZ 130 was making its early test flights, measurements were made to determine the static charge on the ship during flight which would be discharged when ground contact was made. One of the findings was that when the water recovery system was in operation the static charge on the ship was reduced considerably. I do not know the magnitude or extent of the static charge when the water recovery was not operating.

It is quite obvious why the Nazi hierarchy would wish to discount the sabotage theory. Their loss of prestige would have been tremendous, both at home and abroad, unless they could come up with a culprit as they did with the Reichstag fire.[2] But they did not do this in the case of the *Hindenburg* nor was there any evidence released concerning any investigations. One would not expect such releases in Nazi Germany unless it would serve the hierarchy's purpose.

Rosendahl's acceptance of the sabotage theory might be understandable. He was a great believer in the airship and the *Hindenburg* was a ship that had been performing beautifully. It was hard to believe that there was a fatal flaw in the ship that could lead to the static discharge. Furthermore he had done what he could do in giving the *Hindenburg* clearance to land, but according to the static discharge theory, it was the conditions that existed at that time that caused the destruction of the ship. Continuing this line of thought, had the ship's landing been delayed those same conditions would not have existed and the ship would not have been lost. Rosendahl cannot be blamed for the loss: he used excellent judgment in giving landing clearance. But proof of sabotage would have eliminated the possibility that existing conditions had any contributory effect on the loss of the ship.

If the destruction of the *Hindenburg* were an "act of God," that is, the result of an electrical discharge, the liability of the Germans could be eliminated. Whether there were lawsuits against the Germans I do not know, but I have no recollection of any. This could indicate that the Germans would do their best to prove that the loss of the ship was an "act of God." However, the fact that they graphited the lacing cord on the LZ 130, and their strong interest in determining the static charge on the LZ 130, would indicate that they seriously believed the casualty was the result of the discharge of static electricity.

The Nazis could easily have provided fictitious sabotage evidence and "framed" anyone they desired. It would have been a simple matter for them to do this—as they had done in the past. If they believed it to be sabotage but had no evidence, they could have provided it. Since they did not, or chose not to, it would appear that they considered the static discharge theory to be the more reasonable and acceptable and, of course, quite possibly the actual cause of the casualty.

Chapter 16 *Helium Questions and the* Graf Zeppelin II

ollowing the destruction of the *Hindenburg* by a hydrogen fire and the deaths of thirty-five of the ninety-seven persons aboard, Dr. Eckener made the decision that passengers would never again be carried in a hydrogen-inflated airship. Dr. Eckener's statement to me was that there were a number of special conditions that caused the fire and *they all had to occur at the same moment.* He said it was like being dealt four aces in a cold hand of poker (I was surprised that he knew so much about our good American game). It might never happen again and then it might happen the next day.

Immediately after the Lakehurst disaster it seemed certain that the United States would provide the Germans with helium to continue their transatlantic passenger operations. The public and the American Congress, horrified by the gruesome funeral pyre on their doorstep, dramatically presented in all the periodicals and featured nightly in every movie theater, were at one in their resolve to do what they could to prevent any repetition of such a disaster. President Roosevelt himself stated that the destruction of the *Hindenburg* and the loss of life was a horrible catastrophe and that *he* would see to it that the Germans got helium for their airships.

Following the conclusion of the Department of Commerce's investigation into the loss of the *Hindenburg,* to which Dr. Eckener led the German delegation, the Doctor proceeded to Washington to ask for the release of helium to the Zeppelin Company. President Roosevelt, while indicating his favorable attitude, felt it inappropriate for him to see Dr. Eckener when it was up to Congress to amend the Helium Control Act. Taking the hint, the chairman of the Zeppelin Company, through Senator Joseph Robinson of Arkansas, was introduced to Vice-President John Garner, Senator William Borah, the chairman of the Foreign Affairs Committee, and other influential legislators. Garner asked Eckener to testify before the Military Affairs Committee of the Senate—the first time a foreigner had been invited to do so—and Eckener spoke convincingly on the question of whether a helium-filled airship would have military value. The upshot was an amendment to the Helium Control Act providing that the Zeppelin Company should receive a limited amount of helium per year to inflate its commercial airships. The Munitions Control Board, consisting of the Secretary of State, the Secretary of War, the Secretary of the Navy, the Secretary of the Treasury, the Secretary of Commerce, and the Secretary of the Interior, was to ensure that the helium would not be used for military purposes.

In the era of good feeling that followed, Goodyear-Zeppelin's stock rose in Friedrichshafen. As I wrote to President Litchfield on June 25, 1937, "You will be interested to hear that there is now a much better attitude being taken towards cooperation with us in the States concerning the construction of ships than heretofore. I have the impression from Dr. Dürr that he is sincerely interested and is anxiously awaiting the time when we will again get under way with the construction of a ship. It is now being realized that it is more of a problem to build a helium airship than it is to build a hydrogen ship when there is lift to spare."

Dr. Eckener returned home early in June to supervise modifications to the LZ 130, sister ship of the *Hindenburg* then in an advanced state of construction, plus the creation of helium storage facilities, transportation arrangements, and a purification plant. "I have talked with the Doctor several times since he returned [I continued] and one cannot help but get the impression that he is extremely optimistic and he is incidentally in excellent humor. His optimism is not reflected all the way down the line and probably will not be until helium has been definitely released. Everyone realizes over here now that without helium the airship business is finished."

The decision to cease passenger operations with hydrogen meant the end of the old *Graf Zeppelin* as a transoceanic passenger carrier. Coming home from Rio with Captain von Schiller in command when the *Hindenburg* burned, the *Graf* arrived in Friedrichshafen on May 8. Captain von Schiller was prepared to take her out for the next scheduled voyage to South America on May 11, hydrogen and all, but Dr. Eckener canceled the flight over von Schiller's objections. On June 18 the *Graf* was flown to Frankfurt, and was laid up and deflated in the one completed hangar there.

"At present," I continued in my June 25 letter to President Litchfield, "nobody apparently knows whether or not the ship will ever fly again although the general idea seems to be that it will not. It should be possible to build in a few gasoline tanks and water recovery system, ripping out the Blau gas cells, and then to inflate with helium and then to use the ship for training purposes for short flights." Instead, the *Graf Zeppelin* was put on display at Frankfurt, and for fifty pfennings the control car could be viewed from a ramp outside. For one mark the visitor could enter the ship through a loading hatch, walk along the keel and out through the control car. It was often said that the *Graf* made more money this way than she ever did carrying passengers and mail!

At the time of the *Hindenburg* fire, the LZ 130 was approaching completion in the construction shed at Friedrichshafen. Work had commenced on her when the *Hindenburg,* on March 23, 1936, had been transferred from the building shed to the Löwenthal hangar. Completion of the LZ 130 was scheduled for September 1937, and it was advertised that the ship would make her first flight to South America on October 27. In every way the new ship's basic structure was identical to that of the *Hindenburg,* except that 600 mm (23.6 inches) was to be removed from the outer edge of each fin and rudder. With hydrogen inflation, the LZ 130 would have carried seventy passengers as in the modified *Hindenburg,* with extra cabins along the starboard side of the keel in bay 11 abaft the regular passenger accommodations. Other minor improvements were being considered, such as the new ground-adjustable three-bladed propellers and lighter radiators in the power cars. One gas cell was to be made of silk, to test this material which would be lighter than the customary cotton.

On May 6, 1937, the construction of the LZ 130 was far advanced. The basic hull structure was complete and work was being done on the fins and rudders. The engine gondolas were attached to the ship and the first of the Daimler diesels had arrived in Friedrichshafen. Throughout the ship the secondary structure was being completed. The control car needed only the flooring before installation of equipment would begin. The radio room above the control car was taking shape. The structural work on the crew's quarters was complete and the officers' quarters forward of the control car were about 70 percent complete. Most all of the handling line hatches and fittings were in the ship. Cables for the electrical system and the engine telegraphs were being installed. Installation of fuel and water tanks awaited only the placement of the supporting cradles.

Though the work crew responsible for the passenger quarters were working two hours overtime every day, these were far from complete. Actually it was planned to fly the LZ 130 at the earliest possible moment to Löwenthal where the passenger quarters and detail work would be completed. Immediately thereafter, according to plans, work would commence at Friedrichshafen on the next ship, the LZ 131, together with the necessary lengthening of the hangar. Jigs for the first ring, number 92, had already been put down in the ring shed and fabrication of the girders was to begin as soon as a work force was available.

Following the destruction of the *Hindenburg,* all overtime was stopped but work schedules were otherwise maintained. Then immediately after Dr. Eckener's return from the United States, work commenced on altering the LZ 130 for helium operation. Because of the reduced lift with helium (which weighed twice as much as hydrogen and had only 93 percent of the lifting power), weight saving suddenly was of the greatest importance. Continuing in my June 25 letter, I wrote that "there is no question but that the *Hindenburg* was a luxury airship for the ship had such an excess of lift when inflated with hydrogen. The designers certainly were not as careful in holding down the weight as they might have been and that was naturally reflected in the LZ 130. The operators wasted perhaps even more lift than the designers by carrying excessively large operating loads. All of this is ended now and we shall now get an idea as to what these people can do with all their experience in designing and operating when they really have to get down to business."

Some weight would also have to be added, for water recovery was now a necessity in view of the high cost of helium and the limited amounts to be made available. Involved in the design change was not only the water recovery system itself but also additional tankage for recovered water, practically the equivalent of the amount of fuel tankage. In order to maintain the range of the ship, consideration was given to cutting the ship in two amidships and adding an additional bay, but this was not done, partly because the Friedrichshafen shed was too short to accommodate the LZ 130 with a length of 857 feet including the new bay.

The original passenger quarters of the LZ 130 were practically the same as those in the *Hindenburg*. These were completely ripped out and new quarters for only forty passengers were built in. Where the quarters for the *Hindenburg* were on two levels, those for the LZ 130 were essentially on one level. The area that had been the dining room on the port side in the *Hindenburg* was now divided into two rooms, one a lounge and the other a smoking room. Both rooms were of approximately the same size. On the starboard side of the ship, the area that had included the lounge and the writing room on the *Hindenburg* was divided into two equal parts. One section was a lounge similar to the one on the port side, while the remainder comprised four cabins with outside windows. The dining room was elevated (to clear the keel amidships underneath) and extended between the two lounges. Of the twenty double cabins, thirteen had outside windows. This was a tremendous improvement over the inside cabins of the *Hindenburg*. The Luftschiffbau believed that the passenger accommodations of the LZ 130 were not only lighter but also far more attractive than those of the older ship.

After much testing and experimental work, the Luftschiffbau developed a water recovery system capable of 100 percent recovery—that is, recovery of sufficient water to equal the weight of fuel consumed. The power cars were completely rebuilt, retaining only the original engine beds. The cars were made longer, and the diameter increased approximately 30 inches. The propellers were to be tractor instead of pusher as in the *Hindenburg*. All the water recovery equipment was to be mounted within the power car. Each bank of cylinders had its own recovery system so in effect there were to be eight separate units, two in each power car.

The first stage in water recovery was the air cooling of the exhaust manifolds. The exhaust gases were initially at 500 degrees C. The second stage was water cooled; the radiators to cool the water from the second stage were mounted at the aft end of the car. The second stage exhaust gas units were mounted overhead at the top of the car. The last stage was the air-cooled condenser mounted at the front of the car. The gases then passed through the water separator. Cooling of the radiators mounted at the rear of the car was obtained by the use of a blower run by an auxiliary chain drive from the main engine. This blower brought the discharge velocity from the rear of the car up to the air speed of the ship, which reduced the drag of the power car considerably.

Performance of this system was excellent, particularly with a synthetic light oil, Kogasin 2, a by-product of the manufacture of synthetic gasoline from coal. The unusually high hydrogen content (15 percent) was very favorable for water recovery, 100 percent of the weight of the fuel burned

Power car of the LZ 130, the *Graf Zeppelin II,* in assembly. Although the engines were the same as in the *Hindenburg,* the power cars of this ship were radically different from those of the *Hindenburg.* The power cars of LZ 130 were larger, used tractor propellers, and incorporated the water recovery system.

Completed power car of the *Graf Zeppelin II.* The more streamlined shape of the power car made it blend beautifully with the shape of the airship. Note the new three-bladed propeller; made of impregnated wood, it was also ground adjustable.

being obtained at an air temperature of 68 degrees F. A further advantage was the total absence of sulfur in Kogasin 2, eliminating the corrosion problem and deposits on pistons. With a natural fuel oil with a 13 percent hydrogen content, 100 percent recovery was obtained with air temperature of 59 degrees F. Time between overhauls for the radiators was estimated at 1000 hours, the same as for the engines.

To obtain good heat transfer, the soot was periodically blown from the first two stages with compressed air and in the final stages the sludge was washed through using some of the previously recovered and filtered water. It was important to wash the sludge out of the system before it could dry and form scale.

There was no increase in drag of the LZ 130 with water recovery as compared to the LZ 129 *Hindenburg*, and at the same power the LZ 130 would have the same speed as the LZ 129. The only variation was that as a result of the back pressure in the system of 80 mm of mercury, there was a slight increase in fuel consumption; this was minimal, however, amounting to only a few hundred pounds on a North Atlantic crossing.

Dr. Dürr and the operating personnel were well satisfied with the water recovery system and felt that the water recovery problem had now been satisfactorily solved. Although the original projection was for an added weight of 1.1 tons per power car, the final weight was 1.4 tons per power car but this included all the changes made in the power car, structural and otherwise.

The entire electrical plant of the LZ 130, installed as in the LZ 129, was dismantled and completely reworked. The location was changed so that the exhaust gases could be used for cooking in place of the electrical range used on the LZ 129. The heavy truck-type diesels were eliminated and smaller power units with about 30 percent less output were installed.

Some weight was also saved by eliminating all the maneuvering valves and simplifying the valve control system. Manual valving was to be accomplished by manual control of six of the large automatic gas valves—two forward, two amidships, and two aft. Two of the access ladders to the axial corridor were eliminated leaving only one ladder to the center of the ship.

Every effort was made to reduce the weight of the gas cells. The greatest improvement was made in the change to silk instead of cotton for one gas cell. This cell was made up of one ply of silk fabric weighing 40 grams per square meter, and one ply of silk gauze weighing 15 grams per square meter. The various films were applied in eighteen applications, the weight of the finished material being 150 grams per square meter.

Many minor changes were made throughout the ship in an effort to save weight. Savings ranged from as little as one kilogram to very substantial savings in the passenger quarters. The final figure for the weight saved—after including the weight added for the water recovery system—was about 6 tons. Since the water recovery system added 5.7 tons, the weight saved structurally totaled 11.8 tons with an approximate breakdown as follows:

Bow arrangement, lookout, freight rooms, engineers' room, ladders, crew cabins, rudders, cells, valves	4,190 lb
Radiators, piping, spare parts	4,400 lb
Electrical plant, cables, system	2,866 lb
Passenger quarters (40 persons)	12,127 lb
Total	23,582 lb
Less water recovery system	11,466 lb
Net saving	12,117 lb

For the North Atlantic flights, reductions were also planned in the operating loads. These were as follows:

Fuel load, from 62 tons to 49.5 tons	24,255 lb
Lubricating oil	4,400 lb
Water ballast	17,640 lb
Passengers, 10 @ 440 lb each	4,400 lb
Crew, 10 @ 330 lb each	3,300 lb
Total	53,995 lb
Plus net saving in structural weight:	12,117 lb
Total weight saving	66,112 lb
Loss in lift:	
Lift with hydrogen, 7,167,930 cu ft @ 68 lb per 1000 cu ft	485,100 lb
Lift with helium, 7,167,930 cu ft @ 60 lb per 1000 cu ft	429,975 lb
Loss in lift with helium	55,125 lb

With a total weight saving of 66,112 pounds, this would leave available for pressure height at takeoff 11,000 pounds or 2.6 percent to reach an altitude of 1650 feet. Inflation at takeoff would therefore have to be 97.4 percent.

In order to increase the lift at takeoff and then to reach a sufficient pressure height without having to valve helium, the idea of building up artificial superheat in the airship was investigated. (Superheat is the difference between the temperature of the lifting gas within the ship and temperature of the air outside the ship.) Since Frankfurt had an altitude of 330 feet above sea level, for the ship to leave Germany an altitude of 1650 feet would have to be attained.

In September 1937, a series of tests were made at Frankfurt, using cells 12, 13, and 14 of the decommissioned *Graf Zeppelin*. The cells were filled with air and cell 13 was the critical cell, the other two being used primarily to retain the heat in the middle cell. Some tests were made blowing hot air through the ship, but the most successful tests were those made with electrical heating units suspended inside the cells. With 100 kilowatt input to cell 13, a superheat of 32° F was attained after 80 minutes and 41° F after three hours. Another test was made with an input of 68 kilowatts to cell 12 with 14.5° F of superheat attained after 30 minutes, 20° F after 60 minutes, and 29° F after 135 minutes. Further tests indicated that after the power was cut off, the rate of loss of the superheat varied from 0.9° F in 5 minutes and in still air to 1.8° in 5 minutes when air was blown through the ship.

All this testing was done with the idea of conserving helium. The concept of building up artificial superheat appeared practical, but the method of applying it to the ship in actual operation would have to be worked out when she was completed. Since helium never became available, the final application never took place.

It was estimated that an electrical installation to build up artificial superheat in the LZ 130 would weigh approximately 1100 pounds. In order to retain the artificial superheat up to the very moment of takeoff, it was planned to keep the heating cables connected until the ship was in the air and only then to disconnect and drop the cables.

While all these changes were being made—during the latter part of 1937—in the design and construction of the LZ 130, progress was being made at the port of Galveston, Texas, and at the airship base in Frankfurt for the shipment, storage, and purification of the helium that the Germans expected to have released to them by the United States government. The original plan was to accumulate all the high pressure gas cylinders from Friedrichshafen, Frankfurt, Rio, and Pernambuco and ship them to Houston or Galveston where they would be charged with helium. The cylinders would then be shipped in oceangoing vessels, then by barge as far up the Rhine as possible, and finally to Friedrichshafen by rail. These cylinders would be filled to a pressure of 150 atmospheres. A total volume of 6,708,900 cubic feet was considered necessary.

Since the ship was not ready for inflation, this plan was altered to one that was considered to be more permanent. At this time gas in high pressure cylinders was already being transported through the streets of Berlin at pressures of 350 atmospheres. Plans were made to equip two seagoing vessels with high pressure cylinders 33 feet long with gas at 350 atmospheres. These ships would be charged at Galveston. The cylinders would be taken up the Rhine by barge and the gas transferred by pipeline to Frankfurt. The Krupp firm was the only organization in Germany capable of producing these high pressure cylinders.

At Frankfurt storage facilities were being prepared for a total of 7,062,000 cubic feet of helium or sufficient for one complete inflation of the ship. The gasometer was to have a capacity of 494,340 cubic feet and the remainder was to be stored in cylinders at a pressure of 150 atmospheres.

The helium purification plant at Frankfurt was set up to handle 26,500 to 28,250 cubic feet per hour. The general principle to be followed in purifying the gas in any cell was to deflate the cell to 30 percent and to purge this residual 30 percent through the purification plant. The gasometer, with its capacity of 494,340 cubic feet, would take the original 70 percent. These figures were based on the largest cell in the ship which had a capacity of 706,200 cubic feet. Maximum loss of gas expected in the purification process was 6 percent but it was expected that the loss could be held to 2 to 3 percent. The purity to be obtained was expected to range from 99.1 to 99.5 percent.

Thus, approximately six months after the destruction of the *Hindenburg* at Lakehurst, the Luftschiffbau and the Reederei were expecting to stage a comeback with nonflammable helium. A highly disciplined people, accustomed to authoritarian rule first by the kaisers and then by their Führer, the Germans had little comprehension of the American political system with its built-in checks and balances. When the President of the United States made a statement such as that made by President Roosevelt they believed without doubt that helium would be released to them. They had a transatlantic airship almost ready, they were solving the problems of water recovery, and they were prepared to transport the helium from America to Frankfurt where it would be stored and purified. There was every expectation that the LZ 130 would be flying her trials by the middle of April 1938, and that by the summer she would be carrying passengers across the Atlantic with helium.

At the same time, the directors of the Reederei did not have to be told that a ship capable of carrying no more than forty passengers would hardly be profitable, given the higher operating expenses with helium. The next airship, the LZ 131, was to be reworked to carry eighty passengers. This would require a considerable increase in volume, which in theory could be achieved in one of two ways—either by increasing the diameter of the hull as much as possible (by 7½ feet, bringing the maximum hull diameter to 142.6 feet) or by adding extra bays to make the ship longer. Since the first solution would require an entirely new hull design and thus take up two or three years, increasing the diameter was discarded. A lengthened ship, however, could be flying in a year after the LZ 130 left the building hangar, so the LZ 131 was to have the basic *Hindenburg* hull structure with one 15 m bay lengthened to 16.5 m, plus an added 16.5 m bay. The length of the hull thus would be increased by 59 feet, and the overall length would be 862.9 feet. The diameter would remain the same at 135.1 feet. Gas volume with 100 percent inflation would be 7,994,750 cubic feet. Fins, rudders, power plant, and the rest would be as in the LZ 130.

Despite the larger size of the LZ 131, it was expected that the weight of structure would be as much as 9.9 tons lighter than in the *Hindenburg*. There would be a saving of 6.6 tons in the primary structure and 3.3 tons in the secondary structure. The safety factor throughout the design would be reduced from 2.0 to 1.8. A better material was to be used, Duralplat DM 31, resembling the American Alclad. Heat treating would give it 25 percent better strength in compression, though only the channels of the triangular girders were to be so treated. By using soft cables instead of hard-drawn wire for radial bracing in the main ring bulkheads, it was hoped that initial tensions on the wires could be reduced and the ring girders be made lighter. All the gas cells were to be made of silk fabric, while the axial corridor as used in the LZ 129 and LZ 130 would be replaced by a one- or two-cable arrangement.

The passenger quarters were to occupy the bottom of two bays, corresponding to the passenger bay in the two earlier ships plus the next bay aft.

In the forward bay would be the public rooms—lounges along both sides, the dining room across the ship as in the LZ 130, and the smoking room extending athwartships between the port and starboard lounges. On the lower deck of this bay would be the electrical plant, kitchen, officers' mess, and crew's mess. The cabins were all in the after bay and on the same level as the lounge floor. On both sides of the ship would be rows of outside cabins, similar to the outside staterooms of the LZ 130. The remaining cabins would be between the corridor and the keel, but would have outside lighting via light ducts. With the added weight of the passenger accommodations in two bays, the overall net weight saving compared to the *Hindenburg* was still 8.4 tons. The general layout of the LZ 131 was made final in October 1937, and the first bay was to be erected in January 1939. By November 1938, however, it was obvious that helium would not be available to the Germans and all work was stopped. No rings for the LZ 131 were ever erected.

Chapter 17 *Increasing Restrictions as Hitler Prepares for War*

y life in Friedrichshafen in the months following the Lakehurst disaster continued much as before. I was still living expensively at the Kurgarten Hotel simply because there were no decent rooms available anywhere in town. With none of the airships flying, I spent more weekends than usual climbing in the mountains and also did more traveling, particularly to Frankfurt. I continued to enjoy a very close relationship with Dr. Eckener as well as with Knut, though it was at this time that Knut first showed signs of the heart disease that was to cause his death thirty-one years later.

On July 13, 1937, President Litchfield wrote to me to say that he and the officers of Goodyear were doing everything possible in Washington to get helium released to the Germans for the LZ 130, but were encountering considerable opposition from some members of Congress as well as some persons in the Administration. "Litch," as I called him, then wrote in the same letter that since I had been in Friedrichshafen for some time (actually nine months) he would like me to ask Dr. Eckener for permission to send someone else over to replace me, preferably George Lewis, so that I could return to Akron to assist Goodyear-Zeppelin personnel in their research work on airships. By implication, he would also like to see the material I had been accumulating on the Luftschiffbau's research, as I had not been able to send such information home under Air Ministry restrictions.

Dr. Eckener was opposed to such an exchange of personnel because of the current uncertainties. "The government has not yet decided whether or not airships are to be continued," I wrote back on July 27, "and accordingly the Doctor seems to want to avoid bringing up any matter which may be of a questionable nature. A new man coming in here would have to go through the same period of uncertainty that I experienced last fall and at this time it does not seem advisable to inject such a set of circumstances into the picture." I added that possibly I might return to Akron for a month or two without too much of a reaction from the Germans, while I myself would prefer to remain in Friedrichshafen for the time being as there was more activity relating to airships than in Akron.

Some months later, in the fall of 1937, Dr. Eckener suggested that I might make such a brief trip to the United States during November or December, which would be the quiet months, while things would pick up in January and work on gas cells for the LZ 130 would be getting under way at the Ballon-Hüllen-Gesellschaft at Berlin-Tempelhof. Dr. Eckener felt there would be no trouble with the Air Ministry if I returned to Germany after a few weeks, while if someone else came in my place there would be a great many questions asked. It would not even be necessary to report my departure for a short visit, and so I could eliminate the necessity for reporting in on my return.

Furthermore, Dr. Eckener wished that I should take with me to Akron a few square meters of the Germans' new silk gas cell material to be tested and compared with Goodyear gas cell material. On my return he would appreciate my bringing back some Goodyear material together with as much information as possible. Dr. Eckener even went so far as to say that if the foreign exchange were available he would like to purchase a few Goodyear cells for the LZ 130, but of course the Nazis would never have released the funds.

President Litchfield agreed to my making a brief visit, and accordingly I left Germany in December, returning early in January 1938. Before sailing for Germany I stopped at the offices of American Zeppelin Transport Incorporated, where Willy von Meister asked me if I would be willing to represent AZT in Germany, particularly in dealing with AZT's German partner, the Reederei. Mr. Litchfield was agreeable to my acting for AZT provided that it did not interfere with my chief role of representing Goodyear-Zeppelin.

Returning to Friedrichshafen, I stopped in Frankfurt to visit with Max Pruss, and was glad to find him in excellent spirits and greatly interested in what was going on in the airship world. He was still far from well, however, and every day was at the hospital being treated as an outpatient for his severe facial burns. I also went out to the Frankfurt airship base to watch progress on the second hangar, construction of which had started in the middle of 1937. Initially it was scheduled for completion in May 1938, but because of the slowdown in construction of the LZ 130 owing the helium situation, there was a slowdown also in work on the hangar. (It was not until October 1938 that the second hangar was substantially completed and November that the LZ 130, *Graf Zeppelin II,* was transferred there.) Like the first Frankfurt hangar or the so-called *Hindenburg* hangar, the new one measured 902 feet long, 170 feet wide, and 167 feet high, "clear inner dimensions." Unlike the first hangar, the walls of the new hangar were of brick instead of composition, and the floor was wood instead of tile. Wood was held to be a better floor material as it would absorb moisture, keeping the hangar dry and preventing moisture from settling out on the airship.

In February 1938, I was in Berlin for several days, visiting the gas cell plant at Tempelhof and being briefed on the new silk gas cell material by Herr Strobl, who for many years had directed the Ballon-Hüllen-Gesellschaft. On my way back to Friedrichshafen I stopped again in Frankfurt, and found that the sides of the new hangar were being bricked up and slabs being applied to the roof. The foundations for the helium purification plant, compressor house, gas storage and test plant were being laid. I also visited Captain Pruss and discovered that on top of his problems with the burn injuries, he had had his appendix removed.

In the meantime, with the LZ 130 approaching completion, there was concern that helium had not yet been released to the Luftschiffbau, for this meant that the ship could not fly before June. A month later, as I wrote to Mr. Litchfield on April 8:

> There is only one major topic of discussion here and that is helium. As each day rolls around the uncertainty and accompanying unrest grows tremendously and the feeling here now is very nearly the same as it was after the *Hindenburg* casualty and that was bad enough.
>
> There is every indication that the ship will now be ready before the gas is available and since this is evident to the men in design and in the shop the incentive to produce is rapidly dropping off. . . .
>
> What we are all afraid of here is that if the difficulties involved in obtaining helium are too great and if the airship industry becomes too dependent on the American supply that the government will react against the airship, endangering the very existence of the airship industry in this country. It is difficult to estimate how real that danger is, but it is certainly there.

To the astonishment and dismay of all my German friends, who had assumed that helium would be forthcoming as a result of President Roosevelt's promise, it turned out that one member of the Munitions Control Board, Secretary of the Interior Harold L. Ickes, had flatly refused to agree to the sale of helium to the Germans. Since the board had to vote unanimously in favor of the sale, Ickes's stand effectively prevented the deal from going through. He appears to have believed that a helium-filled airship might be used as a bomber in the event of war, and correctly pointed out that naval officers—including Commander Rosendahl—were urging Congress to appropriate funds for naval airships, helium obviously being essential to their carrying out their military mission. Furthermore, Hitler's brutal seizure of Austria on March 12, preceded by a campaign of lies and abuse concerning the situation in that small country, and riots and aggressions by Austrian Nazis obviously directed from Berlin, had outraged American opinion to a degree we could not imagine in Friedrichshafen, and had led to widespread anti-German feeling. Ickes was pleased to find that his refusal to give the Germans helium had the support of many politicians, newspaper editors, and the American people generally.[1]

Although not fully recovered from stomach surgery he had undergone earlier in the year, Dr. Eckener resolved to travel to the United States and plead with President Roosevelt for the release of helium, which was vital to the continuation of the Doctor's life work. Having ascertained through the American ambassador in Berlin that the President would see him, the Doctor sailed in the middle of April. By the time Dr. Eckener arrived in Washington, the helium question had become a major public issue, being debated by the entire American press. Now it was out of the question for President Roosevelt to see him prior to a decision by the full Munitions Control Board. He was able to call on Mr. Ickes on May 14, 1938, for an interview in which the Doctor was unable to disabuse the Secretary of the Interior of the idea that a helium-filled Zeppelin might serve as a bombing aircraft in wartime. Furthermore, Dr. Eckener privately had to agree with

Ickes's underlying reason for refusing the helium: "because your Hitler is preparing for war." Three days later at a meeting of the Munitions Control Board it was voted to suspend the delivery of helium to Germany.

Arrangements were made for Dr. Eckener to meet with the President, obviously to discuss helium, but when the meeting was OK'd word was sent to the Doctor that he was not to discuss helium. Upon meeting the President and exchanging greetings, with Roosevelt asking if the Doctor had had a good crossing, Dr. Eckener immediately asked the question, "Mr. President, when do I get my helium?" I'm quite sure the quotes given here are essentially the same as told to me by Dr. Eckener in June 1938. Roosevelt's reply was that the Americans could not let the Germans have helium because they might use their airships for military purposes. Dr. Eckener stated in his reply that the Germans could not use the airships for military purposes because a big airship would probably be shot down over France long before it could ever get out over the Atlantic Ocean. Roosevelt's comment on this statement was that although the airship might not be of any particular value that way, the Germans still might—or could possibly—try to use it. Dr. Eckener then replied that for the Germans to attempt to fight a modern war with the rigid airships would be somewhat like using men in knights' armor to fight a modern war. To this the President replied that although the Germans would not do this, they could. And Dr. Eckener's reply in turn was that although they could, they wouldn't. It was on this tone that the meeting broke up and the Germans never did get their helium.

Though Dr. Eckener in his memoirs described President Roosevelt as being pleasant and friendly in this interview,[2] he was intensely bitter on his return to Friedrichshafen and felt, in view of Roosevelt's earlier pronouncement, that he had been betrayed. Dr. Eckener always referred to me as "Herr Dick." After telling me of his meeting with Roosevelt, he shook his finger in his characteristic way and said, "Herr Dick, you cannot trust that man. Not once during the meeting did he look me in the eye, and if a man won't look you in the eye, you cannot trust him."

Why was helium not released to the Germans? There may have been several reasons—the political aspects, competition from heavier than air craft, and impending war.

The political reasons were obvious. The *Hindenburg* had been a symbol of Nazi Germany. It carried huge swastikas painted on the vertical fins. Anything that our government might have done to help bring another huge airship with similar huge Nazi insignia into our country could have been unpopular, at least with a segment of the American population. The action taken by Secretary Ickes was simply to take no action and consequently there was no adverse political effect.

The importance of this political effect may not have been very significant. The performance of the *Hindenburg* over the North Atlantic was sensational. A crossing of the North Atlantic was reduced to less than half the time required by the fastest oceangoing vessels. Crossing the Atlantic in the *Hindenburg* was restful and luxurious—a tremendous advance in the age of flight—and was recognized as such. We had already made available to the Germans the Naval Air Station at Lakehurst and there had been no great political repercussions. It would follow that another similar but safer German airship's coming into Lakehurst would have had few political repercussions.

The failure to release helium because of impending war was a reason given later as World War II became more imminent. At the time of the *Hindenburg* casualty, the outbreak of war was approximately two and a half years away. Had helium been made available, the LZ 130 *Graf Zeppelin II* would have been carrying passengers over the North Atlantic in the summer of 1938, one year after the destruction of the *Hindenburg* and one and a half years before war actually broke out.

At no time, from my first contact with the *Hindenburg* in May 1934, and during the construction of the LZ 130 while the Germans anticipated obtaining helium, was there any indication, verbally or structurally, in either the *Hindenburg* or the LZ 130 that either ship would be used for military purposes. The LZ 130 was ultimately used for radar surveillance but this was after the Germans accepted the fact that helium would not be available. The ship was not flown after war broke out. In fact it was scrapped, indicating that the Germans considered it of no value in war.

If the political reasons for not releasing helium to the Germans were not of tremendous importance and the airship had no value to the Germans in war, why was helium not released?

It is my belief that the basic reason helium was not released was because at that time, in the mid 1930s, the airship was so far advanced and ahead of the heavier-than-air competition that the best way for the airlines to eliminate the successful airship program was to ensure that the necessary helium be denied to the Germans. This factor is not admitted or discussed in the literature, and documents on the sabotage of the Reederei's program are hardly likely to be discovered in the archives of Pan American or Imperial Airways. But to me the circumstantial evidence is convincing. Comparing the performance of the airship with heavier-than-air in transoceanic commerce easily shows that heavier-than-air was not even competitive. On the other hand, the airlines were all either government owned or government supported, and could exert heavy political pressure behind the scenes.

Over the South Atlantic, beginning in 1933, Germany's Lufthansa was carrying mail in Dornier *Wal* flying boats from Bathurst in Africa to Natal in Brazil. Since the flying boats lacked the range to make the crossing nonstop, two seagoing vessels specially fitted with large catapults, the *Westfalen* and the *Schwabenland,* were stationed along the route, the one off Bathurst and the other near Fernando de Noronha. Landing on the water, the Dorniers were hoisted aboard with a crane, set on the catapult and refueled, and sent on their way. Although the Reederei was a part of Lufthansa after 1935, there was still an intense internal rivalry, for many German heavier-than-air supporters had never accepted the "gas bag."

The French state airline, Air France, formed in 1933, had absorbed a predecessor company, Aeropostale, which beginning in 1928 had opened an eight-day South Atlantic airmail service from Toulouse to Buenos Aires, but five of the eight days were taken up carrying the mail from Senegal to Natal aboard converted destroyers. Not until January 1936 did Air France open an all-air route with large flying boats carrying mail only. This was a far cry from carrying passengers and mail all the way nonstop from Germany to Pernambuco, as with the *Graf Zeppelin,* or direct to Rio as with the *Hindenburg*.

Over the North Atlantic the *Hindenburg* in 1936 was maintaining regularly scheduled crossings, carrying passengers and mail. Heavier-than-air did not even attempt the difficult crossing of the North Atlantic, with its frequent gales and fog, until 1939 when Pan American Airways and Imperial Airways commenced passenger and mail operations with large flying boats, making numerous refueling stops. Following the end of World War II—approximately ten years after the *Hindenburg*'s success and with the advantage of tremendous wartime technical advances—heavier-than-air, by making intermediate stops in Newfoundland and Scotland, inaugurated a New York to London service. But it was not until 1957, when Pan Am put the DC-7C into service, that heavier-than-air could schedule a direct nonstop service from New York to London, twenty years after the *Hindenburg*'s North Atlantic operations.

Over the Pacific the comparison is mind boggling. While the *Hindenburg* was making regularly scheduled flights over the North and South Atlantic, Pan Am was attempting to island-hop across the Pacific, the longest leg being 2400 statute miles to Hawaii. This operation was not too successful as there were times when the flying boats, heavily laden with fuel, could not get off the water and the passengers would be delayed a few days on one of the remote islands in the Pacific.

By comparison, the *Hindenburg*'s first flight from Germany in the spring of 1936—actually from Friedrichshafen in southern Germany, then northward to avoid flying over France, thence southward and across the South Atlantic to Rio—was a regularly scheduled passenger-carrying flight, nonstop for 5942 miles. This compared favorably with the total distance of 6842 miles across the Pacific from San Francisco to Manila, which Pan Am's Martin China Clippers were flying with four stops.

On that 5942-mile nonstop flight, the *Hindenburg* arrived at Rio with sufficient fuel aboard for another 1040 miles at normal cruising speed. In other words, with added fuel loaded at the expense of some payload, the *Hindenburg* in 1936 could have flown nonstop from San Francisco to Manila! We carried 1.3 tons of mail and express, all the provisions for the return flight, a spare engine weighing approximately 2 tons, thirty-six passengers and fifty-four crewmen, a grand piano in the passenger lounge, and 14 tons of water ballast. Hardly a shoestring operation!

Harold G. Dick and J. A. Sinclair (*left*) during Sinclair's visit from England to Friedrichshafen to see the L Z 130 in summer 1937. (From the collection of Harold G. Dick)

Would the heavier-than-air operators have welcomed this type of competition? It is only reasonable to expect that everything possible would have been done behind the scenes to halt this rivalry, to find some excuse for helium *not* to be released to the Germans. The two obvious rationalizations were based either on the political repercussions or on the possibility of war, with the Germans using the airships for military purposes. Neither reason can be considered valid. I can offer one personal experience tending to prove this argument from the summer of 1937 when the LZ 130 was being remodeled to operate with helium.

When the *Hindenburg* was about to fly early in 1936, I had become well acquainted with Captain J. A. Sinclair, an Englishman who had been involved with lighter-than-air in England. He returned to Friedrichshafen in July 1937, and at Dr. Eckener's request, I showed Captain Sinclair the LZ 130 and explained the changes that were being made so it could operate with helium. He asked me what the Germans were going to do with the ship. I explained that they were going to inflate the ship with helium and put it into service over the North Atlantic.

Captain Sinclair looked at me and said very briefly, "No, they're not. Imperial Airways has already seen to it that they'll not get the helium."

Even in May 1938, before Dr. Eckener's return from the United States, frustration in Friedrichshafen over the denial of helium was leading to the idea of reconverting the LZ 130 to hydrogen. Dr. Eckener on his return recommended that the LZ 130 be again rebuilt to be safe for hydrogen operation. The earliest possible date on which the ship could be ready for inflation would apparently be the beginning of September. "The changes to be made for hydrogen operation are all of a minor nature and consist mostly of the installation of gastight switches, motors, etc.," I wrote on May 18, 1938. "Two months will hardly be required for this work, but Herr Sturm will still require two months to complete his power plant and water recovery installation." On July 12 I wrote to Mr. Litchfield that "over here completion of the LZ 130 is being pushed. This last weekend it was decided that the ship must be ready for hangar tests and for the start of inflation on the 20th of August and ready for flight about the 12–15 of September. . . . Indications are that if helium has not been released the Air Ministry will give the 'go ahead' on hydrogen inflation."

I added, "Personally I cannot see the sense of hydrogen inflation. If the schedule is met there will be only two months available for flight and during those two months the ship will acquire only a few hundred hours' flight time. That will be enough to 'shake down' the ship and will indicate a few minor weaknesses, but it is not enough to really test the water recovery outfit where 500 to 1000 hours will be required to determine how the outfit will stand up. . . . It looks to me as though the advantages to be had are far overbalanced by the risks of hydrogen inflation. If there is no helium available by next spring it might be time to inflate with hydrogen, but I cannot see inflating now."

Actually, as I learned later, it was Dr. Eckener's secret hope that some spectacular demonstration flights by the LZ 130 inflated with hydrogen might lead to the release of helium by the United States government. With the extra lift of hydrogen, ten extra fuel tanks were being installed to give the ship a capacity for 65 tons of fuel, enough for a cruising range, at normal cruising speed of 75 mph, of 9000 miles. Dr. Eckener definitely had a cruise to Greenland in mind, and hoped even to make a demonstration flight to the United States and return nonstop. Alas, the Air Ministry would not give permission for the LZ 130 to be flown outside Germany's borders.

By early August there was great activity at the building hangar. The work crews were putting in sixty-two hours per week, and the flight crew was due to arrive shortly from Frankfurt. It soon developed that they had decidedly mixed feelings about flying with hydrogen. Every one of them had lost friends in the *Hindenburg* fire. Moreover, the very fact that a variety of precautions considered necessary were being taken—a parachute aboard for each man, a cruising altitude higher than formerly, altered landing techniques—did not add to their peace of mind.

On August 16 inflation of the LZ 130 with hydrogen commenced. This was expected to take longer than before as all the hydrogen cylinders had been shipped away, there was no hydrogen in storage, and inflation depended on the daily output of 351,000 cubic feet from the hydrogen plant. Allowing for various delays, it was expected that four weeks would be required for inflation. At this point I was unpleasantly surprised to learn from Dr. Eckener that the Air Ministry had directed that I, as a foreigner, would not be permitted to enter the hangar during the inflation period. To a degree, the new ruling reflected anti-American sentiment resulting from our refusal to give the Germans helium. On the other hand, I was so much a member of the family that the inflation crews were astonished that I was not present as usual, and came by my office to find out what the trouble was!

There was the associated question of whether I could fly in the ship. Mr. Litchfield had written to me on August 23, 1938, that "as you know, I have been definitely opposed to having any Goodyear man fly in a hydrogen-inflated ship, and since the *Hindenburg* disaster I feel even more strongly along that line." Yet I did not want to tell this to my German friends, who might think that I was afraid to share the risks of flying with hydrogen. I was saved the embarrassment when the Air Ministry ruled that I was not to be allowed to take part in any of the LZ 130 test flights.[3]

On September 14, 1938, Dr. Eckener christened the LZ 130 *Graf Zeppelin II* with a bottle of liquid air. He expressed the hope that the new ship, bearing the name of the pioneering LZ 127, would be equally lucky. Immediately after this ceremony, the ship was brought out of the hangar and took off on her first flight, with seventy-four persons aboard. This flight lasted 9 hours 50 minutes. Two of the new three-bladed propellers were on the forward starboard and aft port cars. The other engine cars had the older, four-bladed propellers with their blades shortened to permit higher rpms.

With Dr. Eckener aboard, Captain von Schiller was in command, but on the five subsequent trial flights, he alternated with Sammt and Wittemann. On November 1, 1938, the *Graf Zeppelin II* was transferred to Frankfurt where she occupied the new hangar.

By this time Captain Sammt was permanently in command, and Captain von Schiller shortly left the Luftschiffbau to become director of the airport at Cologne. It is possible that Von Schiller's thinking with respect to hydrogen may have had something to do with his replacement. After the *Hindenburg* fire he did not agree that passengers could be carried only with helium. He advocated keeping the *Graf,* inflated with hydrogen, in service over the South Atlantic. Up to his death in 1977 he maintained that the successful commercial airship had to be inflated with hydrogen. This was probably because of the simpler operation with hydrogen as opposed to helium, with no water recovery system to add weight and complications. Also there was the fact that hydrogen had 10 percent more lift than helium.

I never saw the *Graf Zeppelin II* after her transfer to Frankfurt. Towards the end of her reconstruction in the summer of 1938, I was aware that some kind of secret equipment was being installed on the port side of the passenger accommodations, with the doors and windows boarded up. I felt that it would be undiplomatic to ask questions about this area, which was off limits

Captain Albert Sammt, the last of the German airship commanders. Sammt, Von Schiller, and Wittemann rotated command of the L Z 130 on its early test flights starting in September 1938. On all subsequent flights, Captain Sammt had command of the ship.

The *Graf Zeppelin II* on the mast at Löwenthal. Basically a sister ship to the *Hindenburg*, this ship was altered to operate with helium. Having failed to obtain helium, however, the Germans inflated her with hydrogen.

to me. Not until after World War II did I learn that electronic detection equipment had been installed here by the Luftwaffe Department of Signals, and that beginning in December 1938, after I had left Germany, the *Graf Zeppelin II* had made a whole series of flights over the eastern and western borders of the Reich, and even over England, attempting to gain information on the performance of the radar equipment of Germany's future enemies. On a forty-eight-hour flight up the east coast of England on August 2–4, 1939, the *Graf Zeppelin II* was seen from the ground at several points, and two Spitfires took off from Aberdeen to warn her off. The fact that no radar emissions were detected (the instruments aboard the airship were tuned to the wrong frequencies) caused the Luftwaffe to dismiss British early warning radar as of no operational value, with disastrous results for the Germans in the Battle of Britain.

The *Graf Zeppelin II* made her last spy flight on August 20, 1939, only two weeks before the attack on Poland. She was then hung up and deflated in her hangar at Frankfurt. Here the enmity of the Air Minister, Reichsmarschall Hermann Goering, proved the undoing of both ships named *Graf Zeppelin,* the LZ 127 and the LZ 130: as early as March 1, 1940, he ordered their destruction with the excuse that their duralumin metal was needed for aircraft production. Subsequently the two proud symbols of German aerial supremacy were broken up by a Luftwaffe construction battalion, their own crews refusing to do the deed. On May 6, 1940, with the flimsy pretext that they blocked the takeoff of bombers participating in the great battle in France, both of the Frankfurt hangars, by Goering's direct order, were dynamited. By a melancholy coincidence this was the third anniversary of the *Hindenburg* disaster at Lakehurst.

In spite of the 1936 Air Ministry order prohibiting me from obtaining information on the airships on an official basis, I had continued to enjoy free access to all parts of the airship plant in Friedrichshafen. I could come and go as I pleased in all the hangars, the shops, and even the wind tunnel. As 1938 progressed, however, one restriction after another was put into effect, obviously because in certain areas the work being done had nothing to do with airships. The wind tunnel was the first area put off limits to me, undoubtedly because secret military aircraft designs were being tested. The Luftschiffbau Zeppelin had undertaken subcontracting work for the Dornier firm, but I was not supposed to know that it was constructing wings for the Luftwaffe's new Do 17 twin-engined "flying pencil" bomber. These wings were being fabricated in the airship ring building hangar so I was not allowed in that area. When it was necessary for me to walk through one of these areas where work for Dornier was being done, Knut would accompany me. He would be on the side away from the airplane work and I was supposed not to look around, but to keep my eyes on him while we appeared to be carrying on a very serious conversation.

It was time to leave. The *Graf Zeppelin II* was idle in Frankfurt, but I could not have flown aboard her in any case because of the Air Ministry prohibition. The Munich Crisis of September 1938 convinced the democracies that Hitler was bent on starting a war of conquest in the near future, making it even more unlikely that the American government would release helium to the Germans. In Friedrichshafen and Frankfurt I was increasingly hemmed in by restrictions on where I could go and what I could see. On September 17 Mr. Litchfield had written suggesting that I should plan to spend the Christmas holidays at home with my family, and so I left in late November 1938. I had some idea of returning, but nearly thirty years would pass before I again visited Friedrichshafen.

Chapter 18 *"If Only Nothing Happens to the* Macon*"*

Ostensibly sister firms committed to assisting each other in pursuit of a common goal—the technical development of the big rigid airship and its employment in transoceanic passenger service—Goodyear-Zeppelin and Luftschiffbau Zeppelin did not always work in harmony. While the respect and regard for each other of the two principals, President Litchfield and Dr. Eckener, was unwavering, there were rivalries and jealousies at lower levels. Also there were wide and astonishing differences in the design concepts and structural features of the airships constructed by the two firms, each considering its creation the best. A final difference was that the Goodyear-Zeppelin design showed structural weakness, one of their two big ships being lost through structural failure, while none of the 119 rigid airships built by the Luftschiffbau Zeppelin ever failed structurally because of aerodynamic loads.[1]

The founding of the Goodyear-Zeppelin Corporation had been the result of an agreement reached between Dr. Eckener and President Litchfield. Since Goodyear had no engineering personnel with experience in rigid airship design, it was agreed that certain Luftschiffbau Zeppelin personnel would come to America to design airships for Goodyear. In October 1924, shortly after the completion of the LZ 126 *Los Angeles*, thirteen Luftschiffbau Zeppelin engineers took ship for America and Akron. The leader was Dr. Karl Arnstein, the German firm's chief stress analyst. The others were facetiously called the "twelve apostles" (see page 21). All of these men I got to know, some very well indeed.

Paul K. Helma and Kurt Bauch were stress analysts, Helma being the head stress analyst at Goodyear-Zeppelin. Benjamin Schnitzer headed up structural design, assisted by Walter Mosebach, who in 1924 was quite young and simply wanted to come to the United States. Dr. Wolfgang Klemperer, a brilliant aerodynamicist trained at Aachen under Theodor von Karman, later headed the research department at Goodyear-Zeppelin. Herman Liebert headed up the projects department there. I worked under him from 1930 until I went to Friedrichshafen in the spring of 1934. He was the "bad boy" of the outfit and I think he had more fun than any of them. Lorenz Rieger headed up power plant design at Goodyear-Zeppelin; William Fischer was under him. Eugene Schoettel was responsible for fabric design and gas valves. Eugene Brunner handled patent matters. Erich Hilligardt was an expert on electrical installation, as was Hans Keck on control systems.

Why did these natives of Germany choose to come to the United States? One obvious reason was that with the Versailles Treaty forbidding the construction in Germany of rigid airships larger than 1,000,000 cubic feet and calling for the destruction of the airship building works, these men had reason to believe that they had no professional future in the Fatherland. Another reason was the freedom they expected to enjoy in the United States, whereas in Friedrichshafen they were locked into a rigid hierarchy dominated by Count Zeppelin's associates from the period before World War I. This applied particularly to Dr. Arnstein. A brilliant structural engineer, he had an outstanding reputation as a bridge designer. Some of his railroad bridges in Switzerland and Germany still command admiration today for their efficiency and beauty.

In 1915 Dr. Arnstein had been hired by Count Zeppelin at the urgent request of the German navy to investigate the collapse of the transverse ring structure in some of the early wartime German navy airships. Being successful in explaining the structural failures and advising a solution, Karl Arnstein was made chief stress analyst at the Luftschiffbau Zeppelin. He then developed from theoretical principles a procedure for the stress analysis of the Zeppelin hull framework. This enabled the Luftschiffbau Zeppelin, which previously had enlarged its designs empirically by small increments, to produce in 1916 an airship nearly double the volume of its predecessors. Yet Arnstein must have chafed at being subordinated to the firm's chief designer, Dr. Ludwig Dürr. Although he had been in the Count's employ since 1899, Dürr was a practical man and no match for Arnstein in mathematical skills. Thus, Arnstein must have welcomed the opportunity to run his own show in Akron, and to design rigid airships according to his own innovative ideas. Lastly, as a Jew, Arnstein was discriminated against in Friedrichshafen even before the Hitler period, and must have been doubly glad to have emigrated to America when he saw rampant anti-Semitism all around him during his visits to Germany after 1933.

Following Dr. Arnstein's departure from Friedrichshafen, the chief stress analyst at Luftschiffbau Zeppelin was Arthur Foerster, who himself was still subordinate to Dr. Dürr. Under Dürr and Foerster the company continued to build airships along conservative lines. I have often felt that when the Luftschiffbau Zeppelin management decided who might come to the United States in 1924, they leaned towards releasing those who were advocating changes that the company did not want to make, such as discarding the flat main rings and the cruciform structure supporting the fins in the tail.

In the late 1920s and early 1930s Goodyear-Zeppelin, despite the Depression, seemed to be the more successful partner. Amply funded by the Goodyear Tire and Rubber Company, it began by drawing up plans for a construction shed on the edge of town at the Akron Municipal Airport. This building, which still stands today, was of unusual streamlined shape with "orange peel" doors lying flush to the sides when opened, to minimize eddies around the entrance. Work commenced in the spring of 1929 and the giant "Air Dock," measuring 1175 feet long, 325 feet wide, and 197 feet 6 inches high, "clear inner dimensions," was completed before the end of the year. Meanwhile, working closely with Rear Admiral William A. Moffett, the politically astute and visionary chief of the Navy's Bureau of Aeronautics, and his outstanding engineering subordinates—Commander Garland Fulton of the Construction Corps who headed the Bureau's Lighter Than Air Section, and Charles P. Burgess, a skilled stress analyst—Dr. Arnstein and his design staff had developed plans for a scouting airship that would be the largest in the world. With a streamlined hull 785 feet long and 132.9 feet in diameter, this Design No. 60 would contain 6,500,000 cubic feet of helium when 95 percent inflated, with a gross lift of more than two hundred tons. In 1926, Congress passed Moffett's five-year plan for expanding naval aviation, which included authorization for the construction of two large rigid airships, and in 1928 Goodyear-Zeppelin won the contract for building these two craft, to be designated ZRS 4 and ZRS 5.

The future of Goodyear-Zeppelin at this point looked bright indeed: the U.S. Navy's General Board, a council of admirals advising the Secretary of the Navy on policy matters and shipbuilding programs, had recommended ten such giant airships for scouting in the Pacific, while similar craft could carry forty to one hundred passengers across the Atlantic and Pacific under the aegis of the International Zeppelin Transport Company and the Pacific Zeppelin Transportation Company. By contrast, Luftschiffbau Zeppelin's only operating airship, the *Graf Zeppelin,* completed in 1928, had been financed largely by the contributions of the German people to the Zeppelin-Eckener Fund, and Dr. Eckener had been unable to obtain backing from either private capital or the government for the larger LZ 129. The tables were turned after the Nazi takeover in 1933, when the regime financed the Zeppelin enterprise for reasons of national prestige—and no one in 1928 could have foreseen that the ZRS 4 and ZRS 5 would be the only rigid airships ever built in the air dock at Akron.

To a layman, the Goodyear-Zeppelin design for the U.S. Navy ships later named the *Akron* and the *Macon* and the Luftschiffbau Zeppelin design for the slightly larger *Hindenburg* and *Graf Zeppelin II* would have looked practically identical except for the obvious fact that the helium-inflated American ships concealed their engines inside the streamlined hull, while the power plants in the hydrogen-filled German craft necessarily were housed in engine cars suspended from the hull. Inside the silvery fabric outer covers there were a great many differences. Dr. Arnstein's Goodyear-Zeppelin design presented many innovations which he regarded as improvements, while he was critical of the Luftschiffbau Zeppelin's conservative attitudes and continuation of practices going back to the 1914–1918 period. The differences may be tabulated as follows:

Goodyear-Zeppelin	*Luftschiffbau Zeppelin*
Built-up main frames	Flat main frames
22.5 m main frame spacing	15 m main frame spacing
Rectangular girders	Triangular girders
Three keels	One single keel

Goodyear-Zeppelin	*Luftschiffbau Zeppelin*
No axial member supporting bulkheads	Axial support of bulkheads (corridor)
No cruciform structure at fins	Three cruciform reinforcements of fins
Fins attached to two main frames	Fins attached to three main frames
Rather angular fin contour	Sweeping curve on leading edge of fins

The basic design principles followed by Luftschiffbau Zeppelin are shown above. The *Graf Zeppelin* was built accordingly; and there were no real variations in the design of the *Hindenburg*.

Some of the differences between the Goodyear and Luftschiffbau designs resulted simply from the availability of helium to the U.S. Navy: since helium is nonflammable, the eight Maybach engines of the *Akron* and the *Macon* were mounted inside the hull in *two* load-carrying keels, on the port and starboard sides of the ship, driving propellers mounted on outriggers. A third or access keel ran along the top of the ship beneath the backbone girders.

Instead of being flat, the main frames or rings in the American ships were built up to a triangular section, about 6 feet wide and 6 feet high; hence they were considerably heavier than those in the German ships, but also were inherently stiff and did not require radial and chord wires. (However, an "elastic bulkhead" of wire was installed inside the main rings to restrict for-and-aft surging of the gas cells). Dr. Arnstein had created the built-up main ring, and the Bureau of Aeronautics chose this design in preference to the flat main frames because the triangular space greatly improved accessibility to the gas cells and structure in flight. The Goodyear-Zeppelin ships were built with specially designed rectangular section girders in preference to the Luftschiffbau Zeppelin's conventional triangular girders, the former being considered more efficient from the production point of view. The greater weight of the built-up main frames was compensated for by spacing them 22.5 m apart with *three* intermediate frames, yielding larger gas cells. Abaft the control car in the centerline of the ship was an internal hangar to house five airplanes, in accordance with Bureau of Aeronautics specifications. Since the longitudinal strength of the ship here depended on the keels to port and starboard, the hangar did not weaken the structure.

The differences in design, plus statements from the Goodyear-Zeppelin publicity department extolling the virtues of the *Akron* and *Macon* design, had created professional jealousy at Luftschiffbau Zeppelin and brought about a somewhat strained relationship with respect to the prescribed exchange of engineering information. This had led to my being sent to Friedrichshafen to obtain at first hand information on the LZ 129, the later *Hindenburg,* then under construction.

After the loss of the *Macon* in February 1935, the severest criticism of her design centered around the fins, which were attached to only two main frames, while their angular leading-edge profile produced more concentrated aerodynamic loads at the outer forward edge of the fin than in the

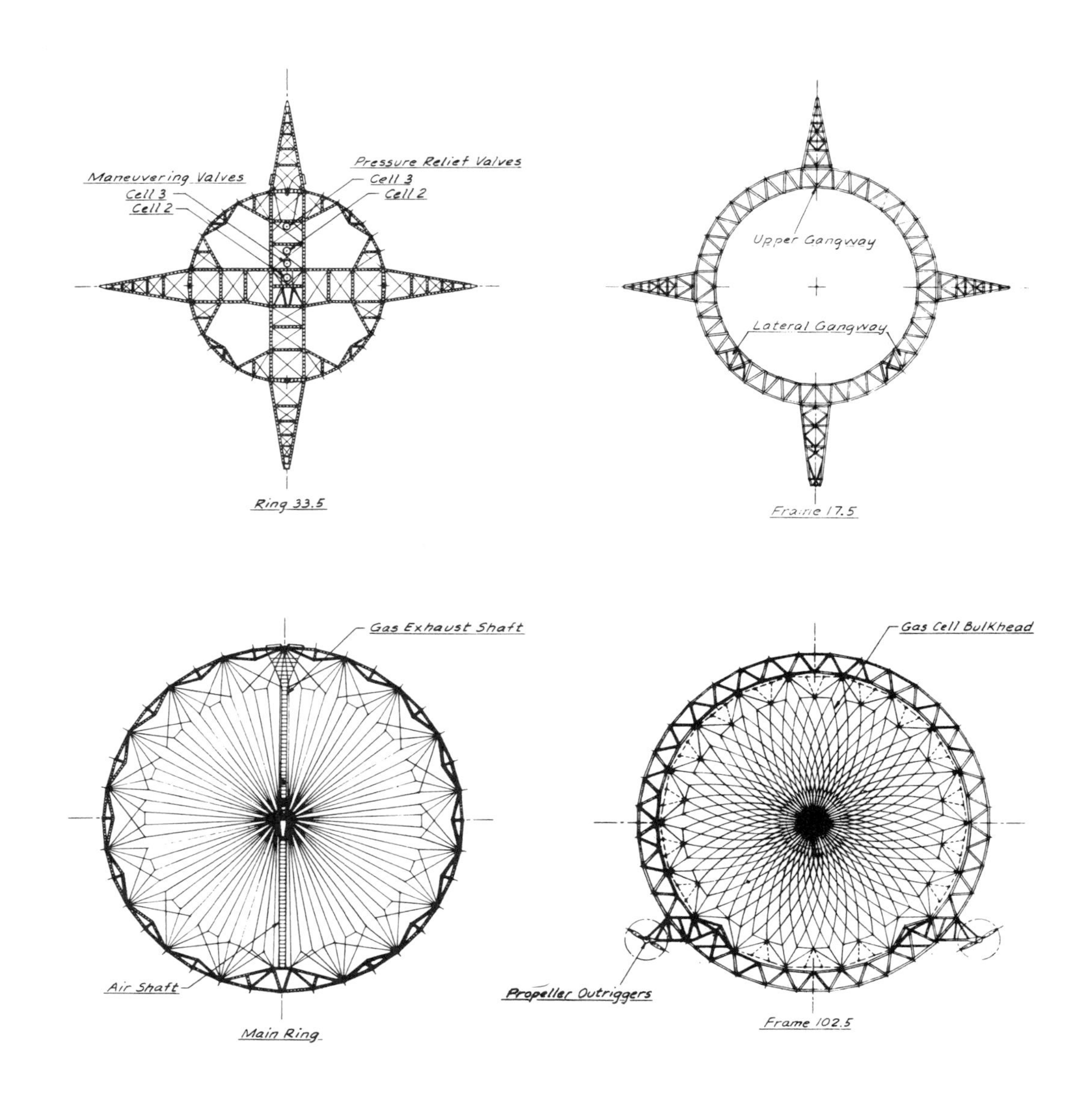

Typical main frame comparison between the *Hindenburg* (LEFT) and the *Akron* or *Macon* (RIGHT). (Drawings by William F. Kerka)

Hindenburg, in which the leading edge of the fins was more rounded. It is only fair to point out that the original design for the *Macon* embodied fins of a much different profile, longer and narrower and attached to three main frames, frame 0 (the rudder post), frame 17.5, and frame 35, while the final design was something of a last-minute improvisation. After the Navy Department had accepted the original design, some of the operators objected on the grounds that during takeoff and landing the lower fin should be visible from the control car. Accordingly the control car was moved aft eight feet and its floor lowered by eleven inches, while the fins were shortened and deepened. The fins were now attached only to main frames 0 and 17.5, and extended forward only to intermediate frame 28.75. In other words, while the leading edge of the deeper fins now had to withstand

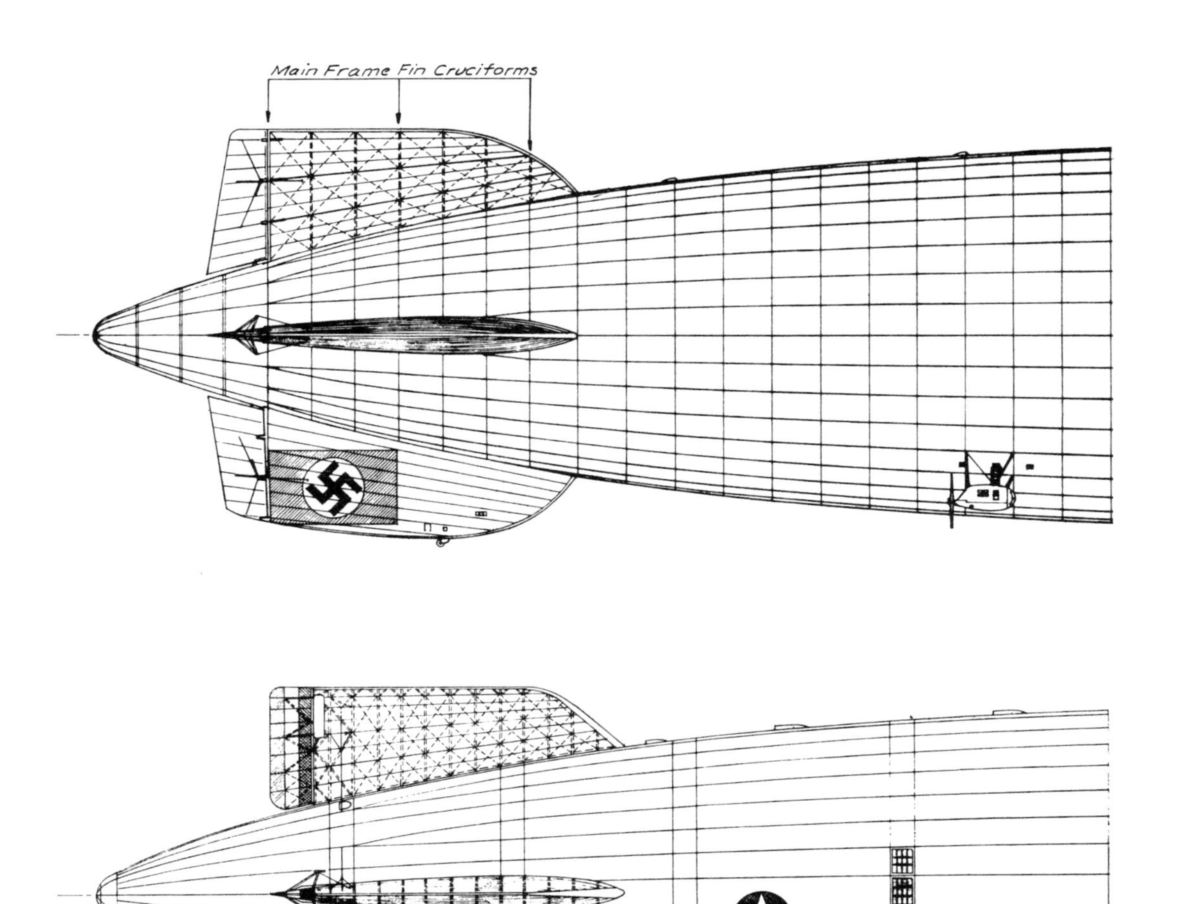

Fin contour and support comparison between the *Hindenburg* and the *Akron* or *Macon*. (Drawings by William F. Kerka)

higher aerodynamic loads, it had no firm point of attachment to the hull.

When the German engineers were brought by President Litchfield to Akron as a design nucleus it could be expected that they would follow standard Luftschiffbau Zeppelin design practices that had been proven in actual airship operations. Since they deviated so far from these standard design practices, it appears to me that they did not carry out the purpose of Goodyear in bringing them to Akron. I seriously doubt whether all these structural changes were brought to Mr. Litchfield's attention. Had they been I believe he would not have approved them unless he was convinced that these changes were improvements over the standard design. The actual performance of the *Akron* and the *Macon* proved that they were not.

Completed in October 1931, the first Goodyear airship, the *Akron*, went down at sea off the New Jersey coast during a thunderstorm on the night of April 3–4, 1933, drowning all but three of the seventy-six people aboard, including Rear Admiral Moffett, chief of the U.S. Navy's Bureau of Aeronautics. While this unprecedented catastrophe spelled the eventual doom of the American rigid airship program, there was no proof of structural failure. The ship was flown into the water, or control was lost and the ship struck the water. On the other hand, the second Goodyear ship, the *Macon,* completed in April 1933, *did* suffer structural failure. On February 12, 1935, while on a scouting flight from the West Coast airship base at Sunnyvale, California, the *Macon* went down at sea after her upper fin tore loose from the hull in flight, causing extensive structural damage aft and deflating the three after gas cells. Fortunately, only two of the eighty-three persons on board were lost.

The first indication to me that the Germans were concerned about the strength of the *Macon* was when George Lewis and I arrived in Friedrichshafen in May 1934. Then and later, a common expression at the German airship works was "if only nothing happens to the *Macon.*" Regardless of professional jealousies between the Goodyear and Luftschiffbau design staffs, Dr. Eckener and his senior assistants knew that in the small world of the great rigid airship, every disaster diminished the prospects for all.

Why were the Germans concerned about the *Macon*? Just before our arrival in Friedrichshafen, the *Macon* had suffered girder damage to main frame 17.5 at the attachment point of the port horizontal fin in severe turbulence over western Texas during a transcontinental flight in April 1934. There had been no structural damage in the German-built LZ 126 *Los Angeles,* or in the LZ 127 *Graf Zeppelin* which on her world flight of 1929 traveling east from Los Angeles to Lakehurst had followed approximately the same route as the *Macon.*

This incident of girder damage to the *Macon* engendered in the Luftschiffbau Zeppelin personnel a lack of confidence in the built-up main frame, and a suspicion that it was underdesigned for the loads imposed on it. Furthermore, the Germans, who certainly knew the assumed design loads for the *Macon* fins, were aware of a sizable discrepancy between the design loads for the upper fins of the *Macon* and those of the *Hindenburg*. The difference was of the order of 8300 kg versus 11,670 kg, with the *Macon* upper fin stressed for only 71 percent of the load assumed for the *Hindenburg* upper fin. Regardless of the type of structure, this would have been adequate grounds for concern even before the *Macon*'s structural difficulty in April 1934. In the end, it would appear that the inadequate design load for the *Macon* upper fin led directly to the structural failure that destroyed her in February 1935.

Why the difference in the assumed design load, and what was the thinking of the rival design teams that led to the difference? At Goodyear-Zeppelin, with the high technical ability of such aerodynamicists as Dr. Klemperer, the selected design loads were based on aerodynamic theory, wind tunnel studies, assumed gust loads, and other calculations. At Luftschiffbau Zeppelin, on the other hand, there was so much practical operating experience that Drs. Dürr and Foerster *knew* the upper fin had to be stressed for higher loads. The Germans as a result of years of flying in all weathers habitually flew *under,* not *through,* fronts and thunderstorms at altitudes as low as 330 feet. Obviously they did so to avoid the turbulence at higher levels, say 660 or 1000 feet. It was also known that the part of the ship subjected to the greatest turbulence was the upper fin, as it would be closer to the more turbulent conditions. Since the Germans stressed the *Hindenburg* fins in 1933 and never changed them, they must have taken this special condition into consideration in requiring a 43 percent higher design load on the upper fin.

After the loss of the *Macon,* there was brought to my attention a series of events which had best be given in their chronological order.

1. As a result of new wind tunnel tests after the ship had been completed, Goodyear-Zeppelin had determined that under certain circumstances the aerodynamic loads on the *Macon* fins, particularly on the forward outboard section, could be such that the safety factor would be reduced to 0.8 (or less than 1.0), indicating that failure would result. This condition would occur if the ship were started into a turn, opposite rudder were applied to stop the turn, and the fin were struck by a gust at the same time. With these findings as the background, Goodyear-Zeppelin agreed to furnish the Navy with the necessary reinforcing items and the Navy agreed to install them in the *Macon.*

At the time of the *Macon* casualty, the Navy had completed the reinforcing work in the vicinity of the lower fin and the two horizontal fins, but not in the area of the upper fin where the structural failure occurred.[2]

2. At one time, when discussing the loss of the *Macon,* Mr. Litchfield stated that he was distressed because he had not been kept fully informed about the structural weakness of the *Macon,* and had he been so informed he would have insisted that the ship *not be flown* until all reinforcing was complete. I can only assume that he knew some reinforcing was required—but that he was *not* advised of the seriousness of the situation. Knowing Mr. Litchfield as well as I did, I believe his reasoning would have been much the same as that practiced by Dr. Eckener: if the future of the airship depended on a decision, you could not *assume* that things would be all right—you had to *know* they would be.

3. In July 1935, Mr. Litchfield, on a trip around the world, arrived at Friedrichshafen to visit Dr. Eckener. Since the *Hindenburg* at that time was still far from completion, I returned to the States with Mr. Litchfield. After I came back to Germany and the Luftschiffbau Zeppelin that October, I was discussing the loss of the *Macon* with Dr. Eckener and he emphatically stated that after studying the weather conditions that existed at the time of the *Macon* casualty, he could find only conditions that the airship should have been able to handle on a routine basis. The inference was quite obvious. The ship was structurally so weak that it had failed under normal operating conditions.

4. A sideline on the omission of the cruciforms in the main frames supporting the fins: Dr. Arthur Foerster, who headed up stress analysis and structural design at Luftschiffbau Zeppelin before coming to the United States after World War II, stated that it was possible for a built-up main frame to handle the loads if properly designed. Obviously the built-up main frames on the *Macon* could not handle the loads, either because of poor structural design or because of wrong load assumptions.

5. My conclusion to all of this information is that the main frames under the fins of the *Macon* were originally underdesigned. This became apparent when the ship encountered turbulent weather and there was girder failure in the main frame supporting the forward portion of the fin. With proper design and load assumptions, there should not have been any failure under these conditions. The ship was structurally weak.

Wind tunnel tests were performed which showed that under certain conditions the actual loads were greater than the design loads, the assumption being that this had caused the girder failure, and did not reflect an inherent weakness in the structure. Arrangements were made to have the main frames reinforced. This reinforcing was done in the way of the horizontal and lower fins. But the upper fin tore loose in an unreinforced area under normal conditions (according to Dr. Eckener), which would indicate that the structure was basically weak.

The question as to whether or not the failure would have occurred had the supporting frames been reinforced with a cruciform structure cannot be positively answered. Design experience would have been on the side of the cruciform: the Germans had successfully used the cruciform structure without failure. The built-up main frames without cruciforms were a relatively new approach and a design error of some type was introduced. Result—failure.

For the more technical minded, I will add a comparison of areas, weights, and maximum design loads for the *Macon* and *Hindenburg* fins and rudders (figures in the metric system):

		Area	Wt	Kg/m²	Load	Load/m²
Upper fin	*Macon*	272.0	1,140	4.19	8,300	30.5
	Hindenburg	291.8	1,430	4.92	11,670	40.0
Horizontal fin	*Macon*	272.0	1,120	4.14	8,300	30.5
	Hindenburg	291.8	1,352	4.55	8,170	28.0
Upper rudder	*Macon*	56.8	352	6.21	2,050	36.1
	Hindenburg	67.2	500	7.43	3,050	45.4

From these figures certain things become apparent:

The design loads and loading per square meter for the horizontal fins in the two ships are within a comparable range. Since the *Hindenburg* with 2.5

kilograms per square meter less than the *Macon* never suffered any failure in the horizontal fins, it would be logical to assume that the design load for the horizontal fins of the *Hindenburg* was correct.

For the upper fin the Germans in designing the *Hindenburg* increased the loading from 28.0 kilograms per square meter to 40.0 kilograms per square meter, or 42.9 percent more. For the *Macon* the loading of the upper fin remained the same as for the horizontal fin. Had it been increased in the same ratio as for the *Hindenburg,* the loading would have been increased from 30.5 to 43.6 kilograms per square meter.

The *Hindenburg* in its entire life-span had no failure in the upper fin, indicating that a loading of 40.0 kilograms per square meter was correct.

The conclusion here is that for the *Macon* the failure to increase the design loads for the upper fin by 30 to 40 percent over the loads for the horizontal fins was disastrous.

The aerodynamic tests made some time before the *Macon* fin failure indicated that the design load of 8300 kilograms for the upper fin was inadequate and that the load for the conditions studied should be 10,375 kilograms. This lower design load would yield a factor of safety of 0.8 for the *Macon* upper fin and failure would result. On the other hand, the resultant loading of 10,375 kilograms, equal to 38.1 kilograms per square meter, would be comparable to the loading of 40.0 kilograms per square meter used in the upper fin of the *Hindenburg*.

The conclusion here is that the Germans were correct in the design loads used for the *Hindenburg* and that Goodyear-Zeppelin was incorrect in the design loads used for the *Macon* upper fin.

It follows that if the *Hindenburg* design for the upper fin was correct and the upper fin should be designed to approximately 50 percent greater unit loading than the horizontal fins, the *Macon*'s upper fin was being routinely subjected to 40 percent higher unit stresses than was the upper fin of the *Hindenburg*. This could have weakened the structure in the area of the *Macon* upper fin to the point where it might prematurely fail even at loads less than the maximum design load.

Consideration of the inadequate design loads may not be the entire story. There is also the possibility that the detail design of the vital critical area was inadequate.

An analysis of the horizontal fins of the *Macon* shows an adjusted weight per square meter of 4.07 compared to 4.55 for the *Hindenburg*. This would indicate a lighter, more efficient design for the *Macon* horizontal fins. It was, however, in the area of the horizontal fins that failure of the main ring girders of the *Macon* took place over Texas in 1934.

The conclusion here is that what appeared to be a more efficient design in the *Macon* horizontal fins, and presumably in the area supporting them, actually was not more efficient but was a weak design that resulted in failure of the ring girders.

The *Macon* had essentially the same weight per square meter for its upper fin as for the horizontal fin, which was 10 percent less on a unit basis than that for the *Hindenburg* horizontal fin, and failure had occurred in the area of the horizontal fin over Texas. It is reasonable to expect the same weakness to have existed in the area supporting the *Macon* upper fin and this would have been aggravated by the fact that the upper fin was subjected to higher loads as shown in the *Hindenburg* design.

Had the *Macon* used the same unit loading as the *Hindenburg*—40 kilograms per square meter—the weight of the upper fin would have been 1495 kilograms or an increase of 355 kilograms over the actual weight, and there would have had to be a similar weight increase in the supporting structure in the range of 30 percent.

It appears to me that there were two basic weaknesses in the *Macon* fin and/or fin area design: *1.* Design loads for the *Macon* upper fin were inadequate: they should have been 30 percent greater. *2.* The detailed design was too light for the loads involved as shown by the fact that the design loads for the horizontal fins were satisfactory (the same as for the *Hindenburg*) but failure had occurred over Texas in the area of the horizontal fins.

Having flown a great deal in the *Hindenburg*—test flights, four South Atlantic crossings, and six North Atlantic crossings—and having had close contact with the operating personnel, I believe the *Hindenburg* was subjected to as severe weather as one could expect. Since there was no structural damage in approximately 5000 hours of flying, it would appear reasonable to assume that the German design loads were correct and that their detailed design was adequate.

Chapter 19 *A Last Look at Friedrichshafen*

December 1938 marked my last actual contact with the Luftschiffbau Zeppelin. After I returned to the States I continued at Goodyear-Zeppelin working on various rigid airship proposals for the U.S. Navy, but it was frustrating for with each proposal a new and slightly altered proposal would be requested. We actually had no hope for the future. I left Goodyear-Zeppelin and transferred to aeronautical sales at the Goodyear Tire and Rubber Company. And in 1941 I married my wife, Evelyn, and we were shortly blessed with two fine children, Hal and Lucy.

During World War II, I had one account—the U.S. Army Air Forces in Dayton, Ohio. Although I had a commission in the Naval Reserve (AVS) I was not called to active duty. Admiral Rosendahl wanted me on active duty at Lakehurst to handle his ideas for developing the big rigid airships and he requested Goodyear to release me for this duty. Since Goodyear and the U.S. Navy (Litchfield and Rosendahl) had agreed not to raid each other's personnel, Litchfield refused to release me on the basis that if there was any development with the big rigids, he wanted me to be with Goodyear.

From time to time during the war years, the newspapers carried accounts of heavy bombing raids on Friedrichshafen by the R.A.F. Bomber Command and by the U.S. Army Air Forces. Actually a great deal of weapons production was going on in Friedrichshafen—radar equipment, jet fighters, and V-2 rocket parts at the Luftschiffbau Zeppelin, tank engines at the Maybach Motorenbau, and tank transmissions at the Zahnradfabrik. The Dornier works at Manzell were busy throughout the war, and towards the end were turning out preproduction versions of the unusual Dornier 335 "push-pull" fighter, which had an engine and propeller at each end of the fuselage and the pilot in the middle. All these factories were legitimate military targets and all were heavily bombed. There was considerable damage to surrounding cities and towns, and the airship works ceased to exist as I had remembered them.

Harold G. Dick.

Dr. Eckener survived the war and in 1947 was brought to Akron where he could spend some time in an airship consulting capacity with the Goodyear Aircraft Corporation. Primarily he was brought to Akron to get him out of Germany during the adjustment period after the war. When I learned that he was in Akron (I had already moved to Wichita) I wrote to him hoping to cheer him up a bit by referring to a ten-day trip we had made together through Switzerland in 1938 when he was celebrating his seventieth birthday. The reply I received to my letter follows:

Akron, Ohio
October 20, 1947

Dear Herr Dick,

I was indeed very happy to receive a letter from you and when I did not answer, it is because I was in New York in the meantime and returned just yesterday. I was very happy and pleased to learn from your letter that you were living in good circumstances and are happily married and are the proud father of two fine children. I should think that is the best fate one can experience in this cursed world, governed mostly (not here of course) by fools and criminals. You seem to me to be suffering only from a little longing for snowclad mountains like the Jungfrau and the Finsteraarhorn? I can understand it because nowhere have I seen more picturesque mountains than in the Swiss Alps. . . . With the greatest pleasure I remember very often the wonderful trip we made together through Switzerland. We Germans are living in remembrances. . . .

Knut is living in Friedrichshafen in two rooms of his house, the main part being occupied by a French officer. He had misfortune: he lost his wife in January of this year. And he is living without the possibility of working, because the Communists fired him—as they did all the managers in the Zeppelin plants. They can do so with agreement and support of the French. Such are conditions in the French zone. You are kind to ask if you can do something for him. Please write him a letter! He will appreciate that very much! and if you will help him a little, then send him a little underwear because he lost most of his in my home where he was living when it burned down (by bombs). My wife and I have been living at our daughter's house in Constance since 1944.

Sammt, Pruss, Wittemann and the others live in Frankfurt. Foerster and Rieger are scientists in Philadelphia, Pa., and Hilligardt will soon follow.

I have been here for six months in a consulting capacity with the Goodyear Aircraft Corp. But I will go back to Constance at the end of this month to my wife who is not in very good health. I had a good time here and am quite recovered of all my troubles. For three weeks I was a guest in Litchfield's home where Mrs. Litchfield fed me very well. Here we had to discuss the problems of "lighter-than-air," whether it would now be obsolete or not. We decided that it is *not*. What the people in Washington are thinking about that is still a mystery.

And now my poor English is exhausted, my dear Harold Dick, and I have to finish by wishing you and your family all the best for all the future.

Sincerely yours,
(signed) Dr. Eckener

Knut's address:
Dipl. Ing. Knut Eckener
Hochstrasse 101
Friedrichshafen (Bodensee)
French Zone (14b)
Germany

This was my last contact with the Doctor. Seven years later he died at his daughter's house in Constance, at the age of 86.

In 1967 when Evelyn and I were traveling through Germany, we visited Friedrichshafen. I was sorry I had done so, for nothing remained of the airship works and the town had been heavily damaged. During one of the bombing raids on Friedrichshafen, the Kurgarten Hotel was set on fire and gutted, the roof being about half burned away. When we arrived, the windows were boarded up, the roof protected where it had been burned; bushes had grown up in the rose gardens, and the lone inhabitant was a watchman patrolling the grounds.

We also had the opportunity to visit with Knut at his home in Constance. Before leaving the States I had written to his last address in Friedrichshafen, but did not receive his reply until reaching Germany:

Dipl.-Ing. Knut Eckener — Konstanz-Staad. Mövenring 16.
4 October 1967

Dear Harold,

Only yesterday did I receive your letter of September 25, 1967, thus I could not send news of myself to Wichita, so I wrote instead to the Frankfurter Hof.

Suddenly to hear again from you after so many years has given me great pleasure. In the meantime I have frequently thought of you. Even though the airship days lie several decades behind us, they remain in memory an interesting and beautiful period of my life.

I have not lived for a long time in Friedrichshafen, but in Constance, Mövenring 16 (Telephone 63793). Unfortunately in the meantime my heart trouble, which I—as you perhaps recall—suffered from for a long time, has gotten markedly worse. I must stay in bed most of the time and cannot be up for more than 2 hours per day. Whether you want to visit me under these circumstances is up to you.

In any case, I have greatly enjoyed hearing once again from you. I hope that you will be able to show to Mrs. Dick Germany from her most attractive side. The season of course is well advanced, for our beautiful Indian Summer is already at an end.

With best wishes to you and Mrs. Dick, I remain,

Your
Knut

When we arrived in Constance I called Knut, we drove out to his house, and Evelyn and I had a delightful visit with him that ran far over the hour that he could be up. After spending over two hours with him reliving some of the things that had happened back in the 1930s, we felt it was absolutely necessary that we leave although Knut wanted to continue reliving those days.

Knut told us of one of the bombing raids on the airship plant. When the raid began he got into his car and drove to the plant where the bombs were being dropped and the cars were being strafed as well. There was a great deal of excitement and it was not until it was all over that Knut found that he also had been strafed and that his car was full of bullet holes—but he had not realized it and had not even been scratched.

Knut expressed concern about the continuation of the Eckener name. His son, who would be the sole survivor with the name, was teaching at one of the universities and up to that time had not married. I was very thankful that we had that visit, for the following spring we learned that in March 1968, Knut had passed away.

And now I have come to the end of the story of my nearly five years at the Zeppelin works, and the unforgettable experience of flying in the *Graf Zeppelin* and the *Hindenburg*.

Is there any place in the future for these great dirigibles? We must remember that the *Hindenburg* was designed and engineered over fifty years ago and that there has been tremendous progress made in the aeronautical sciences, in metallurgy, in instrumentation, and in power plants, none of which has been applied to the big rigid airships that were so far advanced in the 1930s. If such progress were applied to the great dirigibles, as it has been to heavier-than-air, perhaps the great rigid airships with unlimited range would be feasible and, as visualized by Mr. Litchfield and Dr. Eckener, they could again be roaming through the skies.

Appendix A *Glossary of Airship Terms and Technology*

AIRSHIP. Any vehicle that is air supported. The term usually refers to a lighter-than-air vehicle.

DIRIGIBLE. Any lighter-than-air vehicle that has directional control.

FREE BALLOON. A gas or hot air inflated balloon without any directional control.

LIGHTER-THAN-AIR. A vehicle that is supported in air by a gas that is lighter than the surrounding air. Examples: airship, gas-filled balloon, hot-air balloon.

NONRIGID. An airship whose shape is maintained by internal pressure, normally by air-inflated ballonets. Example: the Goodyear blimp.

RIGID AIRSHIP. An airship whose shape is maintained by a rigid structural frame. Examples: *Akron, Macon, Graf Zeppelin, Hindenburg.*

SEMIRIGID. Basically a nonrigid airship but with the addition of a longitudinal keel that carries some of the loads. Examples: *Norge, Roma.*

ZEPPELIN. A rigid airship of the basic structural design developed by Count Zeppelin. Strictly defined, a Zeppelin is a rigid airship built by the Luftschiffbau Zeppelin of Friedrichshafen am Bodensee. The *Graf Zeppelin* and the *Hindenburg* were numbers 127 and 129, respectively, in the company's series.

ZEPPELIN COMPANY. Luftschiffbau Zeppelin.

Technology

Having evolved as early as the year 1916 an efficient airship for the German Navy with a gas volume of 1,949,600 cubic feet, the Luftschiffbau Zeppelin of Friedrichshafen continued building airships with the same design features with little modification except for increases in size. Using the metal duralumin for the girders of the framework, the rigid, streamlined hull was built up of a series of polygonal main frames or main rings (the terms are interchangeable), braced radially by hard-drawn steel wire running from a central fitting to the angles of the polygon, and tied together by longitudinal girders running from bow to stern of the ship. In the bottom of the ship was a keel structure built on three longitudinals constituting an upright V in section with stout lateral bracing at each main frame. Heavy loads were concentrated here, while the keel gave the crew access to the various parts of the ship. The main frames were spaced 15 meters apart, and between each set of main frames was a gas cell, made of fabric and lined with gasproof material—goldbeater's skin until this was supplanted in the 1930s by a film of gelatin latex. There were two unbraced intermediate frames between the main ones, and light longitudinal girders between the heavy main ones, to support the outer cover.

The outer cover was made of light cotton fabric, doped to render it taut and waterproof and to present a smooth outer surface to the air. Stabilizing fins and control surfaces were fitted at the stern, and the engines were slung below the hull in small enclosed cars or gondolas. Early passenger Zeppelins carried their guests in a long external gondola abaft the control car, but in the *Hindenburg* the roomy passenger space was enclosed inside the hull.

The rigid airship was essentially a powered balloon, which was supported in the air by a force equal to the weight of the air it displaced, less the weight of the gas. It was German practice to assume that with standard atmospheric conditions (barometer 760 mm, relative humidity 60 percent, and gas and air temperatures 0° Centigrade), a cubic meter of hydrogen of specific gravity 0.1 would lift a load of 1.16 kg, or 72 pounds per 1000 cubic feet. On this basis the *Hindenburg,* with a gas volume of 7,062,150 cubic feet, had a gross lift of 242.2 tons.

This would not represent the load that could be carried by the airship, as the fixed weights would have to be subtracted—the dead weight of the structure, the girders and wires of the hull frame, the gas cells, outer cover, gondolas, engines, fuel system, crew and passenger accommodations, and other structural features that could not be removed from the ship. In the *Hindenburg* these fixed weights totaled 130.1 tons.

The gross lift, 242.2 tons, less the empty weight, 130.1 tons, yielded the useful lift of 112.1 tons. It was for the airship operator to determine which of several options he would choose in loading the ship. On a long flight much of the useful lift would be devoted to fuel. For a transatlantic journey the *Hindenburg* might load 64 tons of diesel oil, and 3.3 tons of lubricating oil; 20.7 tons of water ballast, stores, and spare parts; crew of fifty with baggage and provisions for 4 1/2 days; 5.5 tons for fifty passengers and baggage, 2.8 tons for their provisions; and 12.7 tons of freight and mail. *This was three years before paying passengers were being flown across the Atlantic in flying boats which made several intermediate stops.*

The lift under standard conditions would vary with changes in temperature, barometric pressure, and relative humidity. In addition, contamination of hydrogen by air diffusing into the gas cells obviously increased the weight of the gas and decreased the airship's lift. With higher air temperatures or lower barometric pressures, the weight of the displaced air was less, and the lift of the airship was less. With low air temperatures or high barometric pressure, the lift was greater. Humidity had a relatively slight effect, though a high air humidity somewhat decreased the lift. In addition, gas temperature relative to that of the air significantly affected the lift. When superheated (usually by the sun's rays), the gas was less dense than it would otherwise be, and the lift was increased; supercooling of the gas (usually late at night) had the opposite effect.

At takeoff and fully loaded, the Zeppelin was "weighed off" and in equilibrium, or sometimes had a few hundred pounds of "free lift." The gas cells were always 100 percent full, or nearly so, at takeoff. Burning fuel made the ship lighter, and her pressure height or ceiling would increase. Since barometric pressure decreased with increasing altitude, the hydrogen would expand, and at a certain relative pressure, spring-loaded automatic valves would open and "blow off" gas until the pressure was equalized. In addition the commander could release water ballast if the ship was heavy owing to loss of gas or a load of rain, snow, or hail. If his craft was light, possibly at the end of a flight after consumption of fuel, or with the gas superheated by the sun, the commander could release hydrogen from manually controlled valves in some of the gas cells. Furthermore, a certain degree of heaviness or lightness could be compensated for by the dynamic lift derived from the thrust of the engines by flying the ship at an angle. Usually the heavy ship was held at altitude by positive dynamic lift developed when being flown nose up, but sometimes gas might be conserved when flying light by flying nose down to develop negative dynamic lift.

Terminology

ADIABATIC HEATING. When a gas is compressed, its temperature rises, owing to the work done on it. (The opposite is true when the gas expands.) With hydrogen the temperature of the gas changes approximately 5 degrees F per 1000 feet of ascent or descent; with helium the adiabatic temperature change is somewhat over 7 degrees F per 1000 feet. During the landing maneuver, adiabatic heating of the gas could make the airship lighter than it otherwise might be.

ALTIMETER. An aneroid instrument, actually a barometer measuring air pressure, this device was graduated to give the altitude in feet. Changes in barometric pressure after takeoff could produce false readings, though corrections could be obtained by radio. A drop of 0.1 inches in barometric pressure could result in the altimeter reading 100 feet too high.

AUTOMATIC VALVES. Spring-loaded valves in each gas cell opened automatically whenever the internal pressure exceeded the external by 7 to 15 mm of water, as when an airship with full gas cells ascended to a higher altitude, expanding the gas.

AXIAL CORRIDOR. A load-carrying member, located approximately on the central longitudinal axis of the ship, which gives access and inspection through the center of the gas cells and reduces deflection of the bulkheads in case of a deflated gas cell.

BALLAST. To enable the airship to ascend to higher altitudes, or to compensate for gas loss or increased loads on the ship owing to rain or ice, water ballast was carried, distributed along the keel in rubberized bags (*Graf Zeppelin*) or metal tanks (*Hindenburg*).

BALLAST BOARD. Indicated the amount of water ballast at various locations throughout the ship.

BALLONET. In nonrigid and semirigid airships, an air-filled compartment inside the main envelope which, being kept under pressure by a blower or other means, maintained a constant pressure in the large bag, regardless of changes in the volume of the gas.

BAY. That portion of the airship separated by two adjacent main frames and containing a single gas cell.

BEAUFORT SCALE. Figures measuring wind velocity at sea, proposed by Admiral Sir Francis Beaufort (1774–1857).

force	0	calm	0–1 mph
	1	light air	2–3
	2	light breeze	4–7
	3	gentle breeze	8–11
	4	moderate breeze	12–16

force	5	fresh breeze	17–21 mph
	6	strong breeze	22–27
	7	moderate gale	28–33
	8	fresh gale	34–40
	9	strong gale	41–48
	10	whole gale	49–56
	11	storm	57–65
	12	hurricane	over 65

BOW TELEGRAPH. Signal for the docking maneuver at the bow.

BULKHEAD. The arrangement of wires or cables bracing diametrically the structure of the main frame.

CONTROL CAR. The "nerve center" of the airship, similar to the "bridge" of a ship, containing all the controls, instrumentation, and signaling devices. Here were the flight stations of the officer of the watch and his navigator. The rudder man handled the rudder wheel in the bow of the car. On the port side of the car was the elevator wheel, with ballast releases and maneuvering valve releases. A chart room and radio room were immediately in the rear of the control car (though in the *Hindenburg* the radio room was above the control car in the keel).

CRUCIFORM. The structure in the plane of the main frames resembling a plus sign (+) connecting the horizontal and vertical fins.

DELAG. Abbreviation for Deutsche Luftschiffahrt Aktien-Gesellschaft, the commercial airship transportation company founded by Count Zeppelin in 1909. Its personnel played an important role in training German Army and Navy flight crews before and during World War I.

DOCKING RAILS. A mechanical ground-handling aid used in conjunction with the traveling mooring mast with the *Graf Zeppelin* and *Hindenburg*. Running through the hangar and for 200 yards out into the field on each side, the rails carried so-called trolleys to which the ship was made fast fore and aft by tackles, so that she was prevented from moving sideways in a wind while entering or leaving the shed.

DOPE: A solution of cellulose nitrate or acetate in acetone, applied onto the outer cover after it was in place, to tauten and waterproof it.

DRIFT: The lateral motion of an aircraft over the ground, caused by the wind blowing at an angle to its course. To steer a true course over the ground, the wind strength and direction had to be determined.

DRIFT METER: Device to measure drift angle and ground speed. A circle marked in degrees, or a series of radiating lines or wires, enabled the navigator to read the drift angle by observing the apparent direction of motion of the ground passing below. Ground speed was obtained by setting the instrument to the altitude shown by the altimeter, and then reading the time required for an object on the ground to move between two transverse lines–or by completing the wind triangles.

DURALUMIN: Name applied to a family of alloys of aluminium with small and varying amounts of copper and traces of magnesium, manganese, iron, and silicon. Its properties were first discovered by Wilm in 1909 and it was first manufactured in Düren, Germany. Being much stronger than the parent metal, duralumin was used for all rigid airship girder work beginning in 1914 in Germany.

DYNAMIC LIFT: The positive (or negative) force on an airship hull, derived from driving her at an angle with the power of her engines. With a large amount of engine power, flying a ship "dynamically" could readily compensate for considerable degrees of heaviness or lightness. At full power and with an angle of 12 degrees, the *Graf Zeppelin* developed a dynamic lift of 12 tons.

ECHOLOT. A nonbarometric altimeter which showed altitude above the surface by measuring the time required for a loud sound to reach the ground and be reflected back.

ELEVATORS. Movable horizontal surfaces at the tail of an airship, attached to the trailing edge of the horizontal fins. Motion upward or downward inclined the ship's nose up or down, and caused her to ascend or descend dynamically.

EMERGENCY BALLAST BAGS. Fabric ballast bags in the nose and tail that could be emptied almost instantaneously.

EMERGENCY CONTROL STAND. An area in the lower fin with elevator and rudder controls to be used in case of failure of the control lines to the control car.

ENGINE CARS. Small streamlined enclosures attached by struts and wires to the hull of the airship, designed to accommodate an engine or engines, and personnel attending them, and to provide enough space to work on the engines in case of a breakdown.

ENGINE TELEGRAPH. The signaling device to the engine mechanic for the desired engine speed or reverse.

FIELD. The area between two rings or frames, either main and intermediate or two intermediate, and two longitudinal girders.

FINS. Vertical and horizontal stabilizing surfaces at the tail of the airship, at the after ends of which were attached the movable control surfaces. Through 1917 the fins were flat with extensive wire bracing. Later, and in the *Graf Zeppelin* and *Hindenburg,* the fins were of thick cantilever construction with a minimum of external bracing.

FIXED WEIGHTS. Total weight of the structure and other permanent installations of an airship. In a rigid airship, this included framework, bracing wires, gas cells, outer cover, gondolas, engines, fuel tanks and piping, ballast sacks, instruments, etc.

FRAME. Ring. These terms are interchangeable.

FREE LIFT. At takeoff, it was German practice to drop about 1000 pounds of water ballast, to give an equivalent ascending force, or "free lift."

GAS CELLS. Filling the entire interior of the airship when 100 percent full of gas, the cells were held in place by wire and cord netting, and made to be both light in weight and as gastight as possible. Through most of the rigid airship era, the only acceptable gasproofing material was goldbeater's skin. Lightweight cotton fabric was lined with goldbeater's skin attached with a special glue, the material weighing only 4.55 ounces per square yard. In 1930, the U.S. Bureau of Standards perfected a synthetic gasproof film of gelatine-latex for the *Akron* and *Macon,* while a similar material was used in the *Hindenburg*. Cells lined with this material were much cheaper and more durable than the earlier types. The weight was 5.3 ounces per square yard.

GAS CELL FULLNESS ALARM. Indicated when the ship is going through pressure height.

GAS CELL NETS. The cord nets that transferred the gas cell pressure to the gas cell wiring.

GAS CELL PRESSURE INDICATOR. Showed whether or not gas was being lost from the cell.

GAS CELL WIRING. The secondary wiring that transferred the gas cell pressure to the structure.

GAS DUCT. The vertical duct that carried the discharged or vented gas from the gas valve to the top of the ship where it was vented outboard.

GAS VALVE. The valve which either automatically or by manual control allowed gas to be discharged from a single cell or interconnected cells.

GAS VALVE HOOD. The streamlined hood on the outside of the ship that vented the discharged gas.

GAS VALVE WHEEL. Made it possible to open the maneuvering valves, usually those through the center of the ship, simultaneously without disturbing the ship's trim.

GIRDER. Transverse rings, and longitudinal members, being required to resist compression and bending loads, were built up of light girders of triangular section. The Zeppelin Company employed drawn duralumin channels connected by stamped lattice pieces. A section of main longitudinal girder, 16.4 feet long, 14.17 inches high, and 10.63 inches wide, weighed only 10.2 pounds but could support a compression load of 4928 pounds. For the *Akron* and *Macon,* Dr. Arnstein designed a four-sided rectangular girder made by riveting together four prestamped strips of duralumin with large lightening holes.

GOLDBEATER'S SKIN. Superb gas tightness, together with light weight, was attained by lining the inside of the gas bags with goldbeater's skin, and this was the preferred gas-proofing material through most of the rigid airship era. Goldbeater's skin was the delicate outer membrane covering the cecum or large intestine of cattle, each animal yielding only one skin measuring not more than 39 × 6 in. The careful handling required in the slaughter-houses, the quantity of skins involved, and the skilled handwork needed in assembling the skins at the gas cell factory, caused the bags so made to be enormously expensive—$14,000 apiece for the cells built by Goodyear for the *Shenandoah.*

GONDOLA. Generic name for any car or enclosure suspended below an airship, possibly derived from the fact that early Zeppelin gondolas not only were shaped like open boats, but also were intended to float on the water.

GONDOLA BUMPERS. One or two were located under each centerline gondola to cushion the shock of landing. These were rubberized air bags enclosed in a rattan framework and covered with heavy canvas. In the *Hindenburg,* however, the bumpers were replaced by low pressure air wheels under the control car and lower fin.

GRAVITY TANKS. Fuel tanks permanently installed over each engine car and feeding the engines by gravity. Machinists' mates of the different cars were responsible for keeping them filled by hand-pumping gasoline from the slip tanks.

GROSS LIFT. The total lift of the gas contained in an airship: equal to the total weight of air displaced minus the weight of the gas.

GROUND CREW. In World War I, German naval airships were walked in and out of their hangars by three or four hundred men. With the larger passenger ships, mechanical aids were used, including docking rails and the traveling mooring mast. With these devices, 350 men were still needed as a ground crew for the *Hindenburg,* 40 of them pulling the traveling mooring mast!

HANGAR. Large buildings at the airship bases, with huge rolling doors at either end. The leeward door was always used for entry and exit except in very light winds. German practice required that the airships be housed in hangars when not in the air, and no mooring-out equipment was used.

HAPAG. Hamburg-Amerikanische Paketenfahrt Aktien-Gesellschaft, the Hamburg-Amerika shipping line.

HELIUM. The second lightest gas known, developed for airship use by the United States, which had a monopoly of its production from natural gas. Helium had the great advantage for airship use of being noninflammable, but it had only 93 percent of the lifting force of hydrogen. In rigid airships, the expense of helium required the installation of heavy condensers to recover water from the engine exhausts to compensate for the weight of fuel burned and to eliminate the need to valve the lifting gas as fuel was consumed.

HULL. The basic structure which gives the airship its rigid classification.

HYDROGEN. The lightest gas known, weighing only 5.61 pounds per 1000 cubic feet in the pure state. In Germany in 1934–38, it was a by-product of the manufacture of oxygen by the decomposition of water electrolytically to produce oxygen and hydrogen. For the *Hindenburg* at Lakehurst, it was produced as a by-product of refining crude oil to produce gasoline and other petroleum products, and was shipped from the nearby Esso refinery at Linden. Hydrogen not only is inflammable, but is also explosive when contaminated by as little as 6 percent of air. The Germans were able to use hydrogen in their airships with relative safety by very strict attention to gas purity. Between flights the gas cells were kept inflated 100 percent full and under slight positive pressure to minimize inward diffusion of air, and purity was checked almost daily; if the purity fell too low, the cell was emptied and filled with fresh hydrogen.

INCLINOMETER. An instrument that informed the elevator man of the up or down angle of the airship; usually a modification of the common spirit level, with a bubble moving in a curved tube with its convex surface upward.

INTERMEDIATE FRAMES. With main frames spaced 15 meters apart, two intermediate frames were spaced at 5-meter intervals between the main frames to reduce bending loads on the longitudinal girders. A ring of girders circled the ship, but there was no transverse wire bracing.

KEEL. A triangular-section corridor running from end to end of a rigid airship, composed of the two bottom longitudinals of the hull and an apex girder. At the bottom of the keel was the catwalk, about a foot wide. (Though there were no hand rails, nobody was ever lost by stepping off it and through the outer cover!) Heavy loads, such as fuel tanks and water-ballast sacks, were hung from sturdy box-girders along the keel, with store rooms for freight and provisions.

LANDING LINES. In making a ground landing, the airship approached the field at a low altitude and dropped from her nose two trail ropes 300 feet long, by which the ground crew hauled her down. Shorter handling lines attached to each main frame along the keel were used to hold her on the ground. So-called spiders—short lines radiating from a snatch block—enabled five or more men to haul on each handling line.

LONGITUDINALS. The main longitudinals were the main lengthwise strength members of the airship. In the *Graf Zeppelin* there were thirteen, triangular in section and running the full length of the ship. The intermediate longitudinals numbered twelve, were lighter, and did not extend all the way to the tail, and were designed primarily to support the outer cover. In the *Hindenburg,* the eighteen main longitudinals and the eighteen intermediate longitudinals were of the same size.

MAIN FRAME OR RING. Major load-carrying frame which carries a bulkhead in its own plane. These were polygons built of girders. In the *Graf Zeppelin* there were ten diamond-shaped trusses in a circle with heavy keel structure at the bottom. The main longitudinals were attached at the tips of the diamond trusses, and the intermediate longitudinals at the middle of the trusses. The main frames of the *Hindenburg* were very similar, but as the ship was larger, there were twelve diamond trusses. The ability of the main frame to withstand the forces exerted by the lifting gas, and by the weight of the ship's structure and loading, was a basic consideration in rigid airship design. Main frames were heavily braced with both radial and chord wires, which also served as bulkheads to prevent fore and aft surging of the gas cells.

MANEUVERING VALVES. Fitted in certain gas cells, these enabled the commander to trim his ship by releasing gas at one end, or on occasion to make the entire ship heavy. The valve pulls were handled by the officer of the watch. (In the *Hindenburg,* for reasons given in the text, the maneuvering valves were in the center of the ship, just above the axial gangway.) When the maneuvering valves were used to make the ship heavy, they were opened all together for a measured interval of time. In the *Graf Zeppelin,* with the toggles of the maneuvering valves of the six midships gas cells connected to the wheel and opened for 90 seconds, the ship was made 1100 pounds heavy.

MAYBACH. In 1909, Count Zeppelin had backed the construction in Friedrichshafen of an engine designed by Carl Maybach especially for airships. The engine was further developed and manufactured by the Maybach Motor Company, a Zeppelin subsidiary, and because of its superior reliability and fuel economy, was used by all German airships, as well as the American *Akron* and *Macon.* The 12-cylinder Maybach engines mentioned in this book had the following characteristics:

year	*type*	*hp*	*rpm*	*wt (lb)*	*wt/ lb/hp*	*fuel consumption gm/ hp/hr*	*Compression ratio*
1924	VL-1	400	1400	2100	5.2	190	5.3/1
1928	VL-2	550	1600	2315	4.2	210	7/1

For the 16-cylinder Daimler-Benz power plants in the *Hindenburg* the corresponding figures were:

year	*type*	*hp*	*rpm*	*wt (lb)*	*wt/lb/hp*	*fuel cons. gm/hp/hr*
1935	LOF 6	1320 takeoff	1650	4348	3.3	
		900 cruise	1480		4.83	170

MOORING MAST. The Germans never used mooring masts until the *Graf Zeppelin* entered service in 1928. The "high" mast was developed in England in 1919, and was copied by the Americans, but no German airship ever moored to the 160-foot high mast at Lakehurst. They employed "stub" masts, with the airship resting on the ground with the rear gondola on a weighted car so the ship could swing around the mast with the wind, at Pernambuco, Seville, and at Los Angeles on the *Graf Zeppelin*'s 1929 world flight. Dr. Eckener preferred to land on the ground and have the ship walked to the mast instead of making a "flying moor." At the home bases (Friedrichshafen, Löwenthal, Frankfurt, and also Rio), there was the traveling mast on rails 20 feet apart, pulled by forty men and adjustable for the heights of the mooring cone of both ships (55.12 feet for the *Graf Zeppelin* and 70.36 feet for the *Hindenburg*).

NOSE CONE. The conical mooring device at the point of the bow where the ship was secured to the mast.

OUTER COVER. The fabric covering of the ship, stretched taut and doped, usually cotton with linen in some areas.

OUTRIGGER. The structural arrangement of struts and cables supporting the power car.

OXYHYDROGEN. A gaseous mixture of oxygen and hydrogen, or loosely, of air and hydrogen—always potentially explosive.

PAYLOAD. The portion of the useful load that earned revenue, devoted to the carriage of paying passengers or cargo or mail.

PITCH INDICATOR. Showed the angle of attack, up or down, of the airship.

PITCHING MOMENT. Tendency, or measure of tendency, to produce motion, especially around a point or axis. It is measured by the product of the force *times* the perpendicular distance to the point or axis. The pitching moment equals force exerted by the center of pressure *times* the distance to the center of gravity. The pitching moment is expressed in foot-pounds forcing the nose of the airship up or down.

PRESSURE HEIGHT. The height at which decreasing atmospheric pressure permitted the hydrogen to expand and build up a relative pressure inside the cells such that the automatic valves opened and gas was "blown off." Following ascent to a preselected pressure height, the commander would ascend or descend at any altitude below this height without fear that gas would be released—an important consideration in flying through thunderstorms.

PROPELLERS. The propellers of the *Graf Zeppelin* and *Hindenburg* were made of laminations of mahogany and walnut. When first completed the *Graf Zeppelin* had two-bladed propellers, later changed to four-bladers. The *Hindenburg* had four-bladed wooden propellers 19.7 feet in diameter. The large propellers were always geared down to move a large mass of air efficiently. The ratio in *Hindenburg* was 2:1. As noted in the text, the LZ 130 *Graf Zeppelin II* had two 3-bladed propellers, adjustable on the ground, the blades of which were machined out of laminations of wood cemented together under great pressure.

RATE OF CLIMB INDICATOR. An instrument that indicated to the elevator man the rate of ascent or descent (in meters per second).

RUDDER. Movable vertical surfaces, at the tail of the airship, whose motion steered the ship to port or starboard.

SHEAR WIRES: Hard-drawn steel wires providing diagonal bracing in all rectangular panels formed by longitudinal girders and transverse rings, and taking the shear loads on the rigid hull.

SLIP TANKS. The ordinary gasoline tanks along both sides of the keel. Some of these tanks could be dropped through the outer cover in an emergency.

SPEAKING TUBE. A tube that carried the voice from the control car to the axial corridor where electrical circuitry was to be avoided because of hydrogen.

STALL. In aircraft, a condition where an excessive nose-up attitude causes a loss of lift, and the aircraft falls out of control. It should be noted that while a "heavy" airship can stall downward like heavier-than-air craft, it can also "stall upward." When flying "light" and nose down, the ship may be inclined downward so far that the dynamic force on the top of the hull diminishes and the excess static lift of the "light" condition will cause it to rise out of control.

STATIC LIFT. The lift of an airship without forward motion, and due solely to the buoyancy of the gas. Contrasts with dynamic lift.

STATOSCOPE. An instrument used by the elevator man to maintain the altitude ordered by the officer of the watch. Very sensitive to barometric pressure, it could be set to the prescribed altitude, and would then register even small variations of altitude. With the statoscope the elevator man was expected to hold his altitude within ±33 feet in good weather, and ±65 feet in more turbulent conditions.

STERN TELEGRAPH. Signal for the docking maneuver at the stern.

STREAMLINING. The shaping of a body so as to cause the least possible disturbance in passing through the air, and hence causing a minimum of resistance, or "drag." Early in World War I, much research on streamlining was done by Paul Jaray in the Zeppelin Company's wind tunnel at Friedrichshafen. Among Jaray's discoveries was that even at relatively low speeds, careful streamlining of all structural protuberances—fins, gondolas, struts, and wires—was important for aerodynamic efficiency of airships.

STUB KEEL (SERVICE KEEL). The short keel close to the engine power car, usually one bay long, carrying necessary service tanks, fuel, lubricating oil, and water.

SUPERCOOLING. A condition (usually obtaining at night) where the gas is cooler than the surrounding air; since the density of the gas is increased, its lifting power is less. As much as −9 degrees F of supercooling has been recorded.

SUPERHEATING. A condition (usually due to the sun's heat being trapped within the hull) where the gas is warmer than the surrounding air. Since the density of the gas is decreased, its lifting power is greater.

THERMOMETER. An air thermometer was in the control car, and also a remote-reading electrical thermometer giving the interior temperature of one of the gas cells. The data provided by these two instruments were essential to determining the lift of the airship, particularly with superheating or supercooling.

TRIM. The attitude of an airship in the air in response to static forces. When weights and lifting forces were properly balanced so that the center of gravity was located directly under the center of lift, the airship was on an even keel and said to be "in trim." If this was not the case, she was "out of trim"; if the nose was inclined downward, "trimmed by the bow," and if the tail was inclined downward, "trimmed by the stern."

TROLLEY. A wheeled truck, pulled by hand and rolling on docking rails. Trolleys served as points of attachment for tackles made fast to the airship fore and aft.

USEFUL LIFT. The amount of lift remaining after subtracting the fixed weights of the airship from the gross lift.

USEFUL LOAD. The load that an airship could carry, equal in weight to the useful lift; included fuel, oil, water ballast, crew, spare parts, passengers, and cargo.

VIBRATING REED ENGINE RPM INDICATOR. A means of checking engine RPM in the control car.

VIBRATING REED SPEED INDICATOR. Showed ship's speed by means of vibrating reeds.

WATER RECOVERY. In U.S. airships filled with helium, to avoid having to valve the scarce and expensive gas as the airship became lighter through consumption of fuel, the exhaust gases from the engines were passed through condensers hung above the gondolas in order to recover the water of combustion. In theory, 145 pounds of water could be recovered from every 100 pounds of gasoline burned. In the *Macon* particularly, more than 100 pounds of water were routinely recovered from 100 pounds of fuel, with the result that the heavy condensers for engines number 3 and number 4 could be removed with a considerable saving in weight. In the *Graf Zeppelin II,* after it was expected she would fly with helium, a much more compact water recovery system was devised to be carried in the engine cars.

WEIGHED OFF. The state of an airship whose lift and load had been adjusted so as to be equal, or whose excess of lift or load had become known by test. (Colloquially, if said to be "weighed off," an airship was in equilibrium, while she would otherwise be said to be "weighed off—pounds heavy," or "weighed off—pounds light.") Before leaving the hangar an airship, through release of water ballast, was weighed off so precisely that one man at each end could lift her off the trestles under the gondolas. In flight, an experienced elevator man could tell a good deal about her static condition from the "feel" of the ship, her inclination nose up or nose down, and the angle of elevator required to keep her at this inclination. Good practice demanded that before landing a ship should be "weighed off" in the air above the field. The engines were stopped or idled and the rise or fall measured on the variometer. In the *Graf Zeppelin,* a rise or fall of 300 feet per minute indicated she was 1100 pounds light or heavy. Gas would then be valved or ballast released to bring her into equilibrium.

WEIGHT EMPTY. See FIXED WEIGHTS.

WIND FORCE. See BEAUFORT SCALE.

Appendix B *Airship Numbering System*

It is common practice in airships to designate a location or station in the ship by a number which usually is the distance from the stern or from a reference point at the stern. As an example, station 100, in the metric system, would be 100 meters forward of the reference point at the stern. Sometimes this reference point would be the main frame at the rudder and elevator hinge points, or it might be the very tip of the stern.

In the case of the *Hindenburg* the stern section was shortened by 2 meters so the stern of the ship which has a diameter of zero is at station 2. This was a change from the original design but all other stations and rings remained unchanged. It means that although the centerline of the mooring cone is at station 247.05 it is actually 245.05 meters from the very stern of the ship.

Major stations are the main frames and on the *Hindenburg* the spacing varied from 13.5 m at the fins to 16.5 m for the four center bays. Other bays are 15 m although the stern bay is 18 m and the bow bay is 11.5 m.

Each bay is numbered starting at the stern and the gas cells carry the same numbers. On the *Hindenburg* the bays and gas cells are numbered 1 through 16 starting at the stern cell.

Location of the intermediate rings follows the same system, as do all the other locations throughout the ship.

Longitudinals are numbered starting at the bottom of the ship with longitudinal number 0 on the lower centerline which is also the catwalk in the lower keel. The numbering progresses to 18 which is on the top centerline of the ship.

Joints follow the above designations and when a joint falls between two longitudinals it will be designated as 1½ if it falls between longitudinals numbers 1 and 2.

Appendix C *Crew Manual of the German Zeppelin Reederei*

No. 166
FOR AIRSHIP CREWS
FOR SERVICE USE ONLY
SERVICE REGULATIONS FOR AIRSHIP CREWS

INTRODUCTION

The German Zeppelin Reederei as the personification of German airship operations stands today at the center of the public's interest. Furthermore, our airships when abroad are representatives of the entire German people. It is clear that the impression made by the German ships and their crews decisively influences the opinions of others concerning the German nation. In order that the airship may be a significant factor in world commerce, the development period must be free from any setbacks. The flourishing German airship operation can justify the confidence which it now enjoys only through safe and on-schedule performance.

Hence every crew member must clearly understand that this task can only be accomplished through the fullest and utmost self sacrifice by the airship crews, and everyone who is considered worthy of belonging to an airship crew must do his duty without question aboard ship, while at the same time, when not on duty, his conduct should reflect credit on the ship and the operating company.

A. GENERAL DIRECTIVES:

1). For personnel in the service of the German Zeppelin Reederei G.m.b.H. (DZR) uniform service attire shall be worn, to make these persons recognizable to the public as personnel of the DZR. This requirement, which applies to all other commercial shipping operations, and through which the individuality of each company may be recognized, has also been shown to be necessary in commercial air activities.

With or without head covering, the salutation will be the German salute.[1]

2). Accordingly the employees of our air company more specifically designated below are required to wear the following specified service uniforms. This requirement is an additional condition of every contract of employment. The uniform regulations will be enforced by the personnel department with respect to every person required to wear uniform.

3). *Service attire:*

a). Blue jacket, double breasted, with 3 pairs of buttons and an accessory pair of buttons. Two lower side pockets and a pocket over the left upper breast; pockets are inset. Long trousers of same material with cuffs.

b). White jacket, single breasted with buttons and turned down collar, 4 patch pockets with buttons. Long white trousers with cuffs.

c). Every employee at his own discretion may have tailored to his order for formal, evening occasions at which it is desirable that he appear in DZR attire, but for which the blue service uniform is insufficient, a so-called mess attire of dark blue fabric, the color and cut resembling the mess attire of the Deutsche Lufthansa. Together with the mess attire there will be worn a fitted vest of blue fabric or white piqué, a high white collar with turned down corners and black bow tie. Vest and jacket with gold DZR buttons.

For tropical wear, the mess jacket may be of the same cut made of white linen or cotton.

d). Service uniform for steward personnel: See Section VI.

4). *Service overcoat:*

a). Summer coat: Of waterproof navy blue material with belt, cut in trench-coat style, double breasted with 4 pairs of buttons.

b). Winter overcoat: Of heavy navy blue fabric with belted back, worn open or buttoned to the top, double breasted with 3 pairs of buttons.

5). *Buttons:*

a). Blue jacket: For flight personnel: Gold DZR buttons, ¾ in. diameter with exception of steward and galley personnel who wear silver buttons. Ground personnel black buttons.

b). White jacket: Flight crew and ground personnel, gold DZR buttons ¾ in., with exception of steward and galley personnel who wear ¾ in. silver DZR buttons.

c). Overcoats:

1). Summer coats: For all employees of the DZR, smooth black DZR buttons.

2). Winter coats: For airship officers, airship engineers, radio officers and ships' doctors, gold DZR buttons ¾ in. All other personnel wear smooth black buttons.

6). *Service caps:*

Navy blue visor caps of doeskin with black mohair band, black patent leather visor and black patent leather chin strap. (For exceptions see B I-IV).

In the period May 1–Sept. 30, as well as in the tropics, the cap is to be worn with a white cap cover.

The cap insignia will be worn by both flight and ground personnel with gold embroidered oak leaves and DZR cap device; stewards and galley personnel will wear silver oak leaves.

7). *Personal linen:*

With the jacket, basically white linen, black necktie. On special occasions, white, stiff turn-down collars, long necktie. Helmsmen, sailmakers, and machinists while on duty may wear a mixed blue and white shirt.

8). *Footgear:*
Basically black shoes. With white clothing white shoes may be worn; on duty inside the ship either footgear with rubber heels, rubber soles, or fabric shoes.

9). *Coverall:*
Gray one-piece coverall of gabardine for duty inside the ship and in engine gondolas.

10). *Leather jacket:*
Helmsmen handling the controls do not wear a coat, but in cold weather will wear a brown leather jacket.

11). *Insignia of branch of service:*

a). Airship officers, airship engineers, radio officers, ship's doctor	With blue jacket above the sleeve stripes: See B I–IV With white jacket: on the shoulder boards above the stripes.
b). All other personnel:	On the upper collar of the jacket in both corners of the collar.

12). *Company insignia, cap badges, DZR buttons:*
On leaving the company's employ, are to be returned to the personnel department, since all insignia are to be worn only by employees of the company and furthermore are protected by law. For cap badges, shoulder boards and buttons charged to the employee, compensation will be paid according to their condition on return. Insignia not charged must also be returned. In case of loss of insignia provided to the employee without cost, charge will be made for replacement as established by the firm.
Loss of insignia which bear serial numbers and are given to the employee only against a receipt is to be reported immediately to the personnel department.

13). *Provision of uniforms:*
All fabric, cap and insignia for service clothing will be provided by the company. In order to ensure the uniform appearance of the clothing, the supply must be by the company.
The choice of tailor to produce the service clothing is up to the individual. However the cut of the clothing must be according to prescribed patterns.
As compensation for procurement and replacement costs, which arise from wearing the service uniform, a yearly uniform allowance will be paid:

1–3 Officers 1–2 Engineers Radio Officer Ship's Doctor Chief Steward	RM 100 = $25.00
4. Officer 3–4 Engineers Galley Personnel Steward, Stewardess	RM 75 = $18.75

Other members of the crew	RM 50 = $12.50
Ground personnel involved in service to passengers	RM 75 = $18.75

14). The service uniform is to be worn aboard ship for duty in the control car, the coverall for inside the ship.
Furthermore, all personnel may wear the service uniform at the base. However, this is *not* for street wear. Exceptions to this are permitted only as follows:
a). To and from duty.
b). When the ship is docked away from the home base and it is not possible to take along civilian clothing.

B. AIRSHIP CREW:

I *Airship Officers:*

1. *Insignia of branch of service for captain through 3. Officer:* Gold colored globe, blue-enameled with silver airship (see Section A 11 a)

2. *Insignia of rank:*

a). Airship captain.	Blue jacket:	4 sleeve stripes, gold, 9/16 in. wide.
	White jacket:	Blue shoulder boards each with 4 stripes, gold, 9/16 in. wide.
	Cap:	Thick gold cord.
b). Airship 1st officer:	Blue jacket:	3 sleeve stripes, gold, 9/16 in. wide.
	White jacket:	Blue shoulder boards, 3 stripes, gold 9/16 in. wide.
	Cap:	Thin gold cord.
c). Airship 2nd officer:	Blue jacket:	2 sleeve stripes, gold, 9/16 in. wide.
	White jacket:	Blue shoulder boards, 2 stripes, gold, 9/16 in. wide.
	Cap:	Thin gold cord.
d). Airship 3rd officer:	Blue jacket:	1 sleeve stripe, gold, 9/16 in. wide.
	White jacket:	Blue shoulder boards, 1 stripe, gold, 9/16 in.
	Cap:	Gold cord interwoven with black.

e). Airship 4th officer:
Insignia of branch of service: Gold stamped steering wheel (See section A 11b)

Insignia of rank:	Blue jacket:	1 narrow sleeve stripe 5/16 in. wide.

II *Airship Engineers:*

1. *Insignia of branch of service for 1st through 3rd engineer:*

		Large gold stamped, cog wheel (see section A 11a)

2. *Insignia of rank:*

a). 1st airship engineer:	Blue jacket:	3 sleeve stripes, gold 9/16 in. wide.
	White jacket:	Blue shoulder boards, 3 gold stripes, 9/16 in. wide.
	Cap:	Thin gold cord.
b). 2nd airship engineer:	Blue jacket:	2 sleeve stripes, gold 9/16 in. wide.
	White jacket:	Blue shoulder boards, 2 gold stripes, 9/16 in. wide.
	Cap:	Thin gold cord.
c). 3rd airship engineer:	Blue jacket:	1 sleeve stripe, gold, 9/16 in. wide.
	White jacket:	Blue shoulder boards, 1 gold stripe, 9/16 in. wide.
	Cap:	Thin gold cord interwoven with black.
d). 4th airship engineer: *Insignia of branch of service:*		
	1. Promoted from engine personnel:	Large gold stamped cog wheel (see section A 11a)
	2. Promoted from electrical personnel:	Gold stamped lightning bolt.
Insignia of rank:		
	Blue jacket:	1 sleeve stripe, gold, 5/16 in. wide.
	White jacket:	Blue shoulder boards, 1 stripe, gold, 5/16 in. wide.
	Cap:	Thin gold cord interwoven with black.

III *Radio Officers:*

1. *Insignia of branch of service:* 2 crossed gold stamped lightning bolts (see Section A 11a)

2. *Insignia of rank:*

a). Chief radio inspector:

Blue jacket:	3 sleeve stripes, gold, 9/16 in. wide.
White jacket:	Blue shoulder boards, 2 gold stripes 9/16 in. wide.
Cap:	Thin gold cord.

b). Radio officers and radio inspectors:

Blue jacket:	1 sleeve stripe, gold, 9/16 in. wide.
White jacket:	Blue shoulder boards, 1 stripe, gold, 9/16 in. wide.
Cap:	Thin gold cord interwoven with black.

IV *Ship's Doctor:*

1. *Insignia of branch of service:* Gold stamped Staff of Aesculapius (see Section A 11a)

2. *Insignia of rank:*

Blue jacket:	1 sleeve stripe, gold, 9/16 in. wide.
White jacket:	Blue shoulder boards, 1 stripe, gold, 9/16 in. wide.
Cap:	Thin gold cord interwoven with black.

V *Helmsmen and Sailmakers:*

Insignia of branch of service:

a). Helmsmen:	Airship chief helmsman:	Large gold stamped steering wheel.
	Airship helmsman:	Small gold stamped steering wheel (see Section A 11b)
b). Sailmakers:	Airship chief sailmaker:	Large gold stamped steering wheel.
	Airship sailmaker:	Small gold stamped steering wheel (see Section A 11b).
c). Engine personnel	Airship chief machinist:	Large gold stamped cog-wheel.
	Airship machinist:	Small gold stamped cog-wheel (see Section A 11b)
d). Electrician:	Chief electrician:	Large gold stamped lightning bolt.

Electrician:	Small gold stamped lightning bolt (see Section A 11b)

e). Personnel in training wear during probation period:

	1). Service uniform as for remainder of crew, but with no insignia of branch of service.
After completing probation period:	2). Airship officer aspirants wear the globe on both sides of the upper collar of the jacket at both corners. Coverall for duty inside the ship.
After completing probation period:	3). Airship engineers wear a large gold cog-wheel on the upper collar of the jacket at both corners.

VI *Steward personnel:*

a). On duty, white high buttoned jacket with silver DZR buttons, blue trousers; in the tropics white trousers are also permitted. The white jackets are to be loose enough that in cold weather warm underwear can be worn under them.
For waiting on table a light blue jacket.

b). Insignia of rank:

1. Chief steward:	Silver braided sleeve stripe.
2. Steward:	No insignia.
3. Stewardess:	Silver gray dress with blue facings.

C. GROUND PERSONNEL:

a). Those employed in passenger service:	Blue jacket, black DZR buttons. Overcoat: Smooth black buttons, cap, insignia for crew.
Cap for station commander:	Thin gold cord.
b). Maintenance group:	No specific service uniform. Work coveralls may be worn.
c). Doorkeepers and guards:	Aviation gray uniform, silver buttons with DZR insignia. "DZR" in metal letters on service cap. Light blue

tabs on collar with insignia of firm.

Watch and Duty Assignments:

Watches The basis for duty aboard the airship is the division of the crew into three watches. The watches are designated in sequence as:

On watch
Off watch
Standby watch (*Pikett-Wache*)

A watch stander who has been relieved is off watch till the next watch is relieved, and then goes on standby watch.

Officers, engineers and radio officers stand 4 hour watches, helmsmen, sailmakers and engine personnel stand 2 hour watches by day (from 0800 to 2000) and 3 hour watches by night (from 2000 to 0800). With the last two groups, it is arranged that the individual helmsman or machinist changes his time on duty and has the same time of going on watch every 4th day.

After brief intermediate landings as in Pernambuco, Rio and Seville, the watch is set on takeoff as if during the layover the watch rotation had continued (example: the watch which had gone *off* duty at 1600 before landing in Pernambuco goes *on* watch next morning with the takeoff at 0600 (see watch bill).

Duty assignments Aside from duty with the steering controls, the gas cells or engines, individual crew members are assigned so-called special functions.

These include:
Maintenance of the ship, its components and instruments.
Weather service.
Cleaning and serving meals.
Handling mail, freight and baggage.
Sanitation service.

Ship's Standing Orders

Smoking prohibited Smoking aboard the ship, and within a radius of 500 feet around the ship, is forbidden.

Foot gear It is forbidden to wear shoes with nails aboard ship. Foot gear with leather soles must have rubber heels. Ordinary heels always have nails and therefore are forbidden. Being in the passenger quarters, including the galley and radio room, is permitted only when on duty.

Cleanliness In using the wash rooms and toilets, every one should be careful to be neat and clean. Use of the open closet right aft is forbidden while flying over land.

Water consumption Since water is available on board only in limited amounts, use water sparingly.

Baggage Each crew member has a free baggage allowance of 33 lb. Heavy leather suitcases are unsuitable for baggage. The following are included in the 33 lb. of baggage:

Underwear
Extra shoes
A second uniform, tropical clothing and cap.
Washing utensils

Fruit or other articles carried on board

Not included in the 33 lb. are:

Leather jacket or overcoat.

Over garments

Customs In accordance with the customs regulations in every country, crew members are forbidden to handle articles for third parties and to carry them in or out of the country aboard the airships.

If anyone wishes to bring presents for foreigners, these articles are to be reported well in advance to the ship's command, so that their being carried will be listed on the ship's manifest. Any charges incurred for customs are to be borne by the crew member.

With every departure and arrival the ship must be officially cleared by customs. Dutiable articles in the possession of the crew must be paid for by each individual. A breach of the customs regulations can lead to the dismissal of the culprit from the crew. Before the ship is cleared by customs all contact with outsiders is forbidden.

Passport Every member of the crew must have in his possession a valid passport. Persons obligated for military service shall receive permission to go abroad from the appropriate military registration office.

Presents Bringing in fruit and other articles is to be limited. The lift of the ship is better utilized for carrying fuel! If each crew member for example brings with him 3 pineapples each weighing 3.3 lb., the total for a crew of 45 men is 440 lb. of pineapples, not counting the bananas and oranges which are always being carried. There is no objection to carrying one pineapple and a few bananas or oranges.

Photography German government regulations forbid carrying photographic apparatus aboard German aircraft. If cameras are carried, these are to be handed over to the ship's command to be carried under lock and key. Aerial photography in foreign countries is permitted only under certain legal restrictions. Permission may be obtained in special cases. As long as such permission has not been conveyed to the ship's command, taking photographs from on board the airship is forbidden.

Health care An officer designated by the captain is responsible for health care aboard the airship. He is assisted by a trained medical corpsman. First aid in case of injury, etc., is provided by this person. Medication can be obtained without cost from the amply stocked ship's pharmacy. Inasmuch as spreading of any disease of bodily function (intestinal disturbances, boils, etc.) can be particularly dangerous in the tropics, everyone is obligated to make use of the remedies carried on board. Only the health officer or the corpsman may dispense medications from the pharmacy.

Order Every crew member must take care that strict order and cleanliness prevail in his quarters. Clothing, linen and other items of private property are to be kept, because of the close quarters on board, in the canvas enclosures or behind the adjacent curtains.

Trash Waste paper, empty bottles, and other trash are to be placed in baskets provided for them, or to be collected for throwing overboard. Throwing such objects directly overboard is forbidden because of the danger that the propellers will be damaged.

Cleaning detail Crew members who from time to time are specially assigned are responsible for good order and cleanliness of the ship. In cleaning detail they are subordinate to the senior helmsman who may be assigned by the ship's command to see to the cleanliness of the ship.

The after part of the ship, up to Ring 155, including the crew's room on the port side (thus excluding the starboard side room for helmsmen and sailmakers) is taken care of by the machinists, while the fore part of the ship forward of Ring 155 plus the room for the helmsmen and sailmakers is cared for by the helmsmen. The more junior crew members are primarily assigned to cleaning detail. For each flight a schedule for cleaning detail will be prepared; the names of machinists are determined by the chief engineer. General ship's cleaning takes place every morning beginning at 0745. In the forward gangway it is the rudder men of the standby watch, in the control car the elevator men of the standby watch. Should the senior helmsman have the standby watch, he relieves the watch at 0745 and leaves the cleaning detail to the junior helmsmen.

In the wash room every man, after washing, should clean up and leave everything in order; the wash basins are to be cleaned out with the available cleaning agents. A separate container is available for used razor blades. In no circumstances are such razor blades to be thrown in the waste paper baskets, since they can cut people emptying the baskets.

Duties of the Elevator Man

After training as rudder men, qualified helmsmen may be employed as elevator men.

The elevator man is expected to maintain the altitude ordered by the officer of the watch. The chief principle in handling the elevators must be to hold the ship at the designated altitude smoothly and with the least possible movement of the elevators, since an excessive pitch-up of the ship caused by the elevators can severely stress the ship herself and create a disturbance in the ship's attitude which will be unpleasant for the passengers.

Angles greater than 5 degrees are strictly to be avoided; at 8 degrees bottles and glasses fall over.

To measure the angle of the ship, 2 inclinometers are available, one reading up to an angle of 5 degrees, the other up to 20 degrees. On the latter the elevator angle can be read off the same as with the indicator.

Before every takeoff the altitude of the landing field is to be set on the barometer. Since the barometers used on board do not show small variations, a statoscope is used for steering with the elevators, which is set to the prescribed altitude and makes possible the detection of a variation of ±65 feet.

The variometers indicate the ascent or descent of the ship in meters per second. Not only in takeoff or landing, but also during flight, they should be watched and any ascent or descent is to be parried with corresponding elevator and angle of the ship.

In good weather the ship is to be held within limits of ± 33 feet, in bad weather within ± 65 feet. Unauthorized rise above pressure height is strictly to be avoided. A report is to be made at once if the ship ascends or descends beyond the limits of the statoscope.

Each elevator man in training will fly with hand control until the watch officer authorizes him to fly with servo control. If the rudder man is steering by the magnetic compass, the elevator man may not use the servo control.

Pressure height, temperature, and the trim of the ship are to be constantly observed and every variation reported to the officer of the watch at once. Together with the rudder man the elevator man keeps lookout forward and to the side. Objects coming into sight are to be reported.

The elevator man should be specially briefed concerning the static and dynamic condition of the ship.

At all times the elevator man should be aware that his duties require a certain degree of independent initiative and he must have a great sense of responsibility.

Before the watch changes the new elevator man will be briefed by his predecessor concerning the condition of the ship and of the weather, as well as objects coming in sight. This information should include the percentage of fullness of the gas cells and the amount of water ballast. Having been briefed, and at night, having become dark-adapted, he reports to the officer of the watch giving the altitude ordered and the condition of the ship. The elevator man going off watch reports with the same information.

Duties of the Rudder Man

The rudder man is required either to hold the ship on the compass course (in degrees) given him by the 3rd officer, or to steer as directed. He will set the designated course on the course indicator.

Every new rudder man is to begin with hand control. He will observe the movements of the rudder as shown on the rudder position indicator on the housing of the gyro compass. In normal weather a rudder angle up to 5° should suffice for course correction. By moving the rudder too frequently the ship will be rendered unstable and will oscillate around the course. In squally weather a rudder angle greater than 15° is permissible only in case of danger. In such an event the rudder man is to call "Rudder is hard over."

The decision as to when servo control may be used is up to the officer of the watch, who must be asked for permission.

While steering with the gyro compass the rudder man will also watch the magnetic compass, and will report if he notices a deviation from the course ordered.

When steering by the magnetic compass, the servo motor is *not* to be turned on, as the electrical current may cause a deviation of the compass.

Furthermore, the rudder man is to maintain a lookout forward and report immediately anything unusual, such as obstacles, heavy squalls, ships coming in sight, etc.

Before relieving the rudder the 3rd officer on watch is to be asked about islands, ships, etc., likely to be sighted.

From his predecessor he will obtain information concerning the course and anything else of importance. Having been briefed and at night, having become dark adapted, he reports on duty to the officer of the watch with the course as ordered, for example "300 degrees by the gyro" or "180 degrees by magnetic." The rudder man being relieved reports to his relief before handing over the rudder any islands, ships, etc., that may be in sight, or such of which he has knowledge that may come in sight; furthermore he reports the weather and after being relieved, gives the course to be steered to the officer of the watch.

Duties of the Sailmaker
The duties of the sailmaker call for the greatest alertness and conscientiousness. The safety of the ship literally depends on the correct maintenance of the gas cells, outer cover, etc.

After thorough instruction by an older man, the junior sailmaker will stand a watch alone and takes on himself a very great responsibility.

At the base: Examine the hydrogen and fuel gas cells, the outer cover, the gas cell netting, the bulkhead wiring and the shear wiring. Rigging for filling the ship: this includes connecting up the hydrogen cells, preparing for purging of the hydrogen manifold and approximate estimate of the percentage of fullness of the cells. (In *Graf Zeppelin* furthermore bringing in the fuel gas manifold and connecting up the fuel gas cells.) During inflation, one has to see to it that the cells are filled at a uniform rate, in order to prevent mistrimming of the ship. After inflation is completed, the cells are disconnected and tied up.

In flight: The landing stations of the sailmakers are as follows: Watch standing in axial corridor, standby watch at stern telephone and off watch in crew's quarters. The watch in the axial corridor makes certain that at takeoff the gas shaft hoods are open, and after venting gas, all automatic valves are to be checked for possible sticking open. The standby watch mans the stern telephone, at takeoff releases the after spider lines and hauls in the after trolly tackles. The sailmaker on watch must familiarize himself immediately after takeoff with the changing state of the cells and the available ballast and drinking water. During his watch the sailmaker will regularly check all bulkhead wiring, the fins and rudders, the entire outer cover and the framework, all gas cells, the automatic valves, the water ballast bags and breeches. Examining the fins and rudders is especially important, particularly in bad weather. Should the ship go to pressure height, the gas cells must be cast loose from the corridor and when descending below pressure height they must again be made fast, to prevent chafing of the cells on the ring girders. Should any kind of defect occur, for example a leaking cell, the outer cover damaged, bulkhead wiring broken, etc., a report must be made at once to the officer of the watch, and he will immediately undertake to repair the damage.

Retrimming with wash water and dirty water during flight may only be done on order of the officer of the watch. Generally the gas shaft hoods are open; should rainy weather be expected, however, these hoods will be closed at once. Whenever gas is being valved, the wires actuating the maneuvering valves must be checked repeatedly in turn; the condition of the cells is to be observed from the axial corridor. Should ballast water be in short supply, rain water will be recovered if the occasion arises. Picking up and distributing this rain water will take place on orders of the officer of the watch.

After the landing the water lines are to be connected up immediately. The dirty water sacks and breeches are to be cleaned and if necessary, filled with fresh water. Simultaneously preparations are made as above for topping up with gas. All this work is to be carried out under the supervision and according to the orders of the officer in charge.

Engineering Personnel
1. *Preparing for flight.*
Test running of the engines will be carried out 45 minutes before the scheduled time of departure.

In the shed, the engines may be operated only with gasoline or diesel oil. During test runs 2 men assigned to each engine as in the watch and station bill must be in the gondola.

During test runs a second fire extinguisher must be ready.

Before starting the engine for the test run, checking of the fuel lines, barricading the shed space under the gondola, making the ship fast to the anchor points in the shed, closing all access hatches, etc., must be carried out.

The engineer on duty will give the signal for starting and stopping the engines for the test run. After the test run, compressed air is to be held at 40 atmospheres and all valves are to be shut. (With *Graf Zeppelin* the ignition is to be turned off only after the engine is stopped.)

2. *During the flight.*
At takeoff, the watch and standby watch personnel designated in the watch and station bill are in the gondola, the off duty watch at the landing stations set down in the station bill. The off duty watch handle the fuel service. This includes distribution, draining water from fuel tanks, checking for flow and overflow, opening and shutting the valves of the fuel tanks, and recording consumption to be entered into the fuel distribution plan.

The watch being relieved which has less than 110 lb. of lubricating oil in the ready-use tank will add a can of oil. If repairs are necessary the standby and off duty watch will assist if necessary.

Cleaning ship and serving food will be carried out according to the watch and station bill. The chief machinists will not be involved in this.

3. *After the landing.*
The duty machinist and the standby remain in the gondola till the ship is secured. The off duty watch will secure the rear gondola on the riding out car.

(In the *Graf Zeppelin* the engine personnel will participate in the "all hands" maneuver of carrying in the gassing manifolds.)

The senior machinist on duty in each gondola is responsible that the engine oil is drained, and that filter cleaning, checking of the engine and propeller are carried out.

The other machinists assist in loading oil and fuel.

(*Graf Zeppelin*: At the mooring mast, two machinists designated in the watch plan have duty at the mast, the remainder according to weather conditions in the gondolas or at the riding out car.)

In case of unusual major repairs, *all* machinists are to stay with the ship until the senior engineer on duty has made up a duty schedule for the repair job.

Weigh off and landing maneuver
The call for landing stations (haul up or weigh off) will be given through the ship by signalling long, short, short, long. All hands will proceed immediately and as rapidly as possible to their stations as given in the landing stations bill. They are to remain at their stations until the maneuver is completed. When ordered to move through the ship to trim it they will take their new stations as promptly as possible. Running back and forth in the ship during the maneuver without orders from the ship's command is strictly forbidden. "Secure from landing stations" is announced by the signal long, long, long. *On no account may the landing station be left before then!*

LANDING STATIONS

(Note: GZ refers to *Graf Zeppelin*)

	On Watch	Standby Watch	Off Watch
Watch officers	Control car	Check elevator man and ballast board	Standing by in navigation room
3rd officers	Log book	Forward landing wheel (GZ telephone in navigation room)	Engine telegraphs
Radio officer	Radio room	Radio room	Telephone in crew room
Elevator man	At the elevators	Bow	Handling lines above the control car (GZ navigation room)
Rudder man	At the rudder	Bow mooring cable (GZ crew room)	Stern lines (GZ bow)
Sailmaker	Speaking tube in axial corridor	Axial corridor aft (GZ telephone in stern)	Crew room
Chief engineer	Stern position (GZ telephone in engine gondola)		
2–4 engineers	Engine gondola #1 (GZ engine gondola #3)	Engineers' room (GZ in charge in crew room)	Engine gondola #4
Machinist Gondola #1	In engine car	In engine car	After landing wheel (GZ gangway over Gondola #1)
Machinist Gondola #2	In engine car	In engine car	Crew room (GZ gangway Ring 95)
Machinist Gondola #3	In engine car	In engine car	Crew room
Machinist Gondola #4	In engine car	In engine car	Crew room (GZ gangway Ring 125)
Machinist Gondola #5	In engine car	In engine car	Crew room
Training duties	In gangway	Engineers' room	Crew room
Electrician	Generator room	Generator room	Crew room
Stewards	In the passenger quarters		
Cooks	In the galley (GZ one cook in the crew room)		
Doctor	In the passenger quarters		
Special duties	Personnel in training, works personnel, always in crew room. Any change requires the previous agreement of the ship's command		

Topping up with gas at the mast
Rigging for gassing up is an all hands maneuver in which every crew member must take part. This operation consists of bringing aboard the gas manifolds and connecting up the gas and water lines at the masthead.

The time at which the all hands maneuver will terminate is to be set by the officer in charge of gassing up.

After conclusion of the maneuver, and during the topping up with gas, one sailmaker and one steersman will remain *in the ship* and two machinists will remain on watch *at the mast head.* The duration of the watch shall be 4 hours.

For topping up with water ballast, 2 people will be specially assigned.

The assignment to watches will be by the officer in charge of gassing up. A schedule of watch assignments, signed by the First Officer, will hang in the navigation room.

The gassing up duty does not exempt individual crew members from taking part.

Weather service
Ashore. During the airship's layover periods at the home base, all helmsmen as well as those in training will participate in the weather service.

As with the weather service aboard ship, the personnel assigned to the different watches prepare and dictate reports.

The navigation officers will give personal instruction to crew members in training concerning the use of weather codes, their deciphering and use in drawing weather maps. Writing up the morning weather map must take place in time to deliver it by 11 A.M.

On the day of departure, the maps must be drawn as quickly as possible since the time of takeoff and the call for the ground handling crew depends on the weather.

The following arrangement for the duty personnel concerned with the weather service ashore will prevail: the watch following the one which on the homeward journey has prepared the last weather map will after landing start with the first weather map requiring to be drawn. The next map ashore, whether the morning or the evening map, will automatically be prepared by the following watch. To avoid overloading of individuals, every one is obligated strictly to follow this arrangement.

The 3rd officer involved in the practical supervision of ship's activities is relieved from the weather service while at the base.

Meals
Since the airship's galley must simultaneously provide for passengers and ship's crew, the times designated for the crew to fetch their meals are strictly to be adhered to.

Fetching meals: Certain persons will be detailed for fetching food, who will pick up food at the serving counter in the prescribed containers. The cook will be advised as to when and for how many persons the food is to be fetched. After the meal the eating utensils are to be cleaned and stored in the space provided for them. The galley will be directed not to provide food outside of the stipulated meal times. Unless matters are otherwise arranged with the galley, meals will be served at the following times: Breakfast 7.30 and 8.10, lunch 11.15 and 12.10, afternoon coffee 15.30 and 16.10, and dinner 18.30 and 20.10.

Since aboard the *Graf Zeppelin* it is very noisy in the forward part of the ship, arrangements with the galley will be made only via speaking tube and will be limited to the most necessary matters. In the red tent[2] one of the junior helmsmen, sailmakers or electricians will be assigned to fetch meals, set and clean off the tables, and clean up afterwards. Personnel assigned this duty will be 3rd elevator man, 3rd rudder man, 2nd and 3rd sailmakers and 3rd electrician. Those involved will draw up their own duty plan so that the work is evenly divided.

Generally meals will be guaranteed on board only during the flight. At bases outside Europe, after landing, one will go to the canteen for meals. In case of doubt the ship's command will decide.

Any problems, wishes or complaints will not be directed to the cook, but to the food committee. This committee is made up of one member each of the seaman and technical branches together with a representative of the officers' mess. For every member of the committee, a deputy will be named from the crew.

Special functions[3]

1st officer: Has the responsibility for directing the maintenance of the entire ship including gas cells. Is in charge of the watch and station bill.

1st 2nd officer: Overall navigation, communications systems, controls and ballast system.

2nd 2nd officer: Clearances by officials; communication with agencies, personnel officer. Responsible for mail and cargo service.

1st 3rd officer: Under the supervision of the 1st officer, in charge of rigging for gassing and gassing up ship, in rotation with the 2nd and 3rd 3rd officers. Handles ship's papers, weather service according to watch bill.

2nd 3rd officer: Responsible for charts, flight reports, weather service according to watch bill.

3rd 3rd officer: Providing and maintaining navigational instruments, tables, statistics, weather service according to watch bill.

Radio officer: Handling the mail, also as necessary making lists of passengers and freight under supervision of 2nd 2nd officer.

1st elevator man: Rigging for gassing, and gassing up the ship. Maintenance and servicing of the built-in instruments (altimeter, barographs, statoscope, clocks).

2nd elevator man: Physical handling of the mail and freight services (Postal drops, etc.). At the disposition of the 3rd officers for preparing for navigational procedures and rigging the control car. Weather service according to watch bill.

3rd elevator man: Assists with mail and freight. Keeps all landing equipment in order (landing lines, spiders, trolley tackles, grommets, etc.). Cleaning duties in the control car. Weather service according to watch bill.

1st rudder man: Maintenance of crew's living and sleeping quarters. Ship's equipment, supervises cleaning ship. Maintenance of mooring equipment. Weather service according to watch bill.

2nd rudder man: Maintenance of ballast and gas valve control wires, as well as fresh and dirty water lines. Cleaning duties in ship. Weather service according to watch bill.

3rd rudder man: Maintenance of rudder and elevator control lines, maintenance of toilet facilities. Assists with mail and freight. Cleaning ship. Weather service according to watch bill.

1st sailmaker: Maintenance of gas cells, valves, outer cover, netting and gassing up equipment.
2nd sailmaker: Maintenance of the ballast bags and breeches, patching material and spare fabric for the gas cells, outer cover, netting and ballast system. At the disposition of the 1st sailmaker.
3rd sailmaker: Assists 1st sailmaker.

When making repairs, installing replacement parts, making alterations or correcting problems in the course of the above special functions, personnel entrusted with these responsibilities will constantly remain in contact with their designated superiors.

(Aboard the airship *Graf Zeppelin,* duties in the "red tent" will be performed by the 3rd elevator man, 2nd and 3rd rudder men, 2nd and 3rd sailmakers and 3rd electrician.)

Explanation of special functions

1st elevator man: The 1st elevator man is in charge during rigging for gassing and during gassing or topping up of the ship. Personnel working under him at this time will obey his orders. The 1st elevator man will inquire of his superiors well in advance for necessary instructions concerning special procedures or deviations from standing orders. In addition he maintains the built in instruments such as altimeter, variometer, statoscope, and clocks.
2nd elevator man: He oversees the actual handling of mail and cargo. Takes custody of mail and cargo before takeoff. (3rd elevator man and 3rd sailmaker assist him.) Lists the number of letters and their weight. Second officer on duty will be informed that receipt of mail is in order. During the flight he handles the mail in cooperation with the assigned radio man. Preparations for mail drops and more particularly, handing over the mail, is in the hands of the second elevator man. It is very important that the mail be correctly delivered to the different destinations. Before takeoff the control car is to be rigged for flight. In agreement with the 3rd officer on watch, custody of the navigational instruments. Weather service according to watch bill.
3rd elevator man: Assists 2nd elevator man with handling the mail and cargo service, 3rd elevator man primarily responsible for correct delivery of cargo, as 2nd elevator man is occupied with mail. The cargo is delivered according to the manifest, with the recipient signing a receipt. Furthermore he is responsible for the efficient state of the landing equipment (mooring ropes, spiders, trolley tackles, etc.) Weather service according to watch bill.
1st rudder man: Responsible for state of crew's living and sleeping spaces. Has custody of ship's equipment and supervises cleaning of ship. Furthermore the mooring equipment is to be kept in good working order. Weather service according to watch bill.
2nd rudder man: Regularly checks the ballast and gas valve control wires. Has custody, under supervision of the health officer, of the ship's pharmacy and as necessary, holds sick call. Maintains fresh and dirty water systems and their valves and pumps in efficient condition. Weather service according to watch bill.
3rd rudder man: Maintenance and regular inspection of rudder and elevator control cables. Furthermore, the toilets are to be kept in good working order. In

case of need, helps with receipt and delivery of mail and cargo. (Since the watch may vary when landing, he should get in touch with 2nd elevator man beforehand.) Weather service according to watch bill.

1st sailmaker: Primarily responsible for maintenance of the gas cells, gas valves, outer cover, netting and gassing arrangements. 2nd and 3rd sailmakers are at his disposition.

2nd sailmaker: Maintenance of ballast sacks and breeches. Makes certain that at all times there is sufficient patching material and extra fabric for cells, outer cover, netting and ballast system on board. (Otherwise at disposition of 1st sailmaker.)

3rd sailmaker: To assist 1st sailmaker.

Graf Zeppelin: Fetching meals and cleaning the "red tent" is the responsibility of the 3rd elevator man, 2nd and 3rd rudder men, 2nd and 3rd sailmakers, and 3rd electrician.

The necessary preparations, for example cleaning the spaces, washing eating utensils and setting tables, etc., are to be carried out sufficiently promptly that the scheduled times can be met. Furthermore care will be taken that a full water container is always on the foremost small girder. Concerning times for eating see "mealtimes."

Appendix D *The Speyer Airship Project*

The Speyer airship project, according to Douglas Robinson, is unknown to present-day airship historians, and accordingly I have agreed to Doug's suggestion that mention be made here of this enterprise.

I first learned of the operation late in 1935 through a news item in the Friedrichshafen *Seeblatt,* which related that a Herr Brinkmann planned to construct ten pressure airships of 706,200 cubic foot volume to be ferried to the United States and used for nighttime advertising flights. Speyer, being a town of about 35,000 inhabitants on the Rhine south of Mannheim, was not exactly next door to Friedrichshafen, and it was not until January 11, 1936, that I was able to make my first visit there. I found the entire enterprise, the Deutsche Luftfahrzeug Gesellschaft (German Aircraft Corporation), operated by a Herr Otto Brinkmann, with Engineer Nikolaus Basenach as his chief designer. The premises were composed of a former aircraft plant built late in World War I, but all the equipment was new. The presence of Basenach convinced me that the enterprise was a serious one, for this gentleman had been associated from 1907 to 1914 with Major Gross of the Prussian Army Airship Battalion in building a whole series of prototype semi-rigid airships, the last one, the M IV of 688,540 cubic feet, having served in the German Naval Airship Division in 1914–15. Basenach as a designer was held in high respect by personnel of the Zeppelin Company.

One ship was under construction at the time of my first visit, the car and fins being built at Speyer, while the bag was being manufactured at the famous August Riedinger Balloon Factory at Augsburg. An erecting shed was under construction at Speyer. As soon as the first ship was ready to fly (supposedly by late spring), the second, third, and fourth ships would be started, though my own impression after going around the works was that the first flight would not take place till late summer of 1936.

While the Brinkmann design was classified as a semi-rigid airship, it was really a non-rigid, there being no keel. In order to eliminate any possibility of the car, 125 feet in length, acting as such, it was divided into three sections, the sections being connected together with flexible rubber link couplings. Dimensions of the ship were:

Volume	741,510 cu ft
Length	318.3 ft
Diameter	71.2 ft
Overall height	91.8 ft
Maximum width	72.2 ft

There were to be three gasoline-fueled Junkers L 5 engines of 365 HP each located in the car, driving propellers on outriggers, the two forward ones being reversible. The rubberized fabric bag would have had three separate compartments for hydrogen, with three ballonets. The fins, of welded steel tubing, would be attached to two cruciform girders running through the bag, one at the rudder post and the other at the leading edge, so that the fins would still stand in their normal relationship with the bag deflated. The advertising display was to consist of three rows of seventeen letters each on each side of the bag, each letter being 7 ft 10 in high and being made up of 131 light bulbs. The letters could be detached and the ship then could carry sixty passengers for short-haul operations. Gross lift was calculated at 50,700 pounds and useful lift at 27,500 pounds. While the Brinkmann airship presented some features of Basenach's M IV, the shape of the bag was to approximate as closely as possible that of the small passenger Zeppelin *Bodensee* of 1919, which was of approximately the same volume. This would have been quite a large pressure airship, but of course it had to have the range to be flown across the Atlantic to the United States via Seville, the Azores, and Bermuda.

Herr Brinkmann was living quite high off the hog during the time I knew him in Speyer. On one occasion I drove with him in his supercharged Mercedes on the Autobahn from Speyer to his home (classified as a "villa") on the hills overlooking the River Neckar near Heidelberg—a beautiful place where I spent the night with his family. During the drive on the Autobahn (where to this day there is no speed limit) it was raining quite heavily and we passed everything as though they were standing still—at 100 MPH or more! I thought we would take off and start flying at any moment! Later, Brinkmann and his wife accompanied us on the *Hindenburg*'s second flight to Lakehurst. Arriving there on the morning of May 20, 1936, they were whisked off to New York, apparently to confer with Brinkmann's financial associates, and returned in time for the *Hindenburg*'s departure for Frankfurt that evening, having been only sixteen hours in the United States!

I spent a lot of time with Brinkmann during my visits to Speyer trying to discover what his operation was all about, and realized rather early that the primary purpose was to get money out of Germany. (The Nazi government, because of their shortage of foreign exchange, would not allow money earned in Germany to be taken out of the country.) Brinkmann, who was quite a promoter, had lined up as backers some people in New York who held considerable sums in blocked marks which they could not get out of Germany, and had founded the Deutsche Luftfahrzeug Gesellschaft to build airships with the blocked marks, and then fly them to the States. The ships would then be sold to his American firm, "Sky Ads Corporation," and his backers in New York would have gotten at least some of their money out of Germany. Brinkmann wanted me to join his firm to fly the ten ships across the Atlantic, but I preferred to stick with Goodyear-Zeppelin and its more certain future.

The Nazi government apparently realized what Brinkmann was trying to do with the blocked marks, and placed increasing obstacles in the way of releasing them to the Deutsche Luftfahrzeug Gesellschaft. My last visit to Speyer in December, 1936, found everything at a standstill for lack of money. Only two arches remained to be erected for the hangar, and about half the roof was covered in. The bag for the first airship, completed a year earlier, was still in Augsburg. The car was only partially complete, not much farther along than in the spring of 1936, with engines and a few instruments installed. A lot of work had been done on the signs to go on the bag. Shortly afterwards the company closed its doors. There was talk of selling the first ship to the Reederei for training, but the airship operators argued that even a large non-rigid was not suitable for training personnel to fly in the big ships of the *Hindenburg* type. Herr Basenach wrote to me personally to ask if Goodyear would take over the operation: I had to reply by saying that some years earlier Goodyear had closed down all operations in Germany because of the impossibility of getting their profits out of the country. The Speyer airship never flew, while some time later I learned that Herr Brinkmann had been arrested and imprisoned on charges of defrauding the Fatherland through his currency manipulations.

Notes

Introduction

1. Hugo Eckener, *Im Zeppelin über Länder und Meere* (Flensburg: Verlagshaus Christian Wolff, 1949), p. 486.
2. Ibid.

Chapter 1

1. This first Goodyear airship was the *Akron* of 375,000 cubic feet capacity, built for Melvin Vaniman for a transatlantic flight. During a test ascent on July 2, 1912, at Atlantic City, the *Akron* exploded, killing Vaniman and his crew of four.
2. President Litchfield's German is slightly in error. The *Misthaufen* (manure pile) in the center of the courtyard is the main feature of the traditional three-sided German *Bauernhof,* or farm building.

Chapter 2

1. By contrast, no other country succeeded in creating a transatlantic passenger operation with airships. The British R 100 which flew to Canada in 1930, and the R 101 which crashed and burned later that year in France on a projected flight to India, had palatial passenger accommodations but never carried paying customers in regular service. It was only in 1939 that one could fly the Atlantic commercially by airplane, when Pan American Airways and British Overseas Airways Corporation started service with flying boats with numerous refueling stops.

Chapter 3

1. The Santa Cruz base was completed late in 1935. The hangar still stands today, serving the Brazilian Air Force.

Chapter 5

1. In 1934, I was twenty-seven years old and Knut was thirty-two.

Chapter 6

1. Henry Cord Meyer and Stephen V. Gallup, "France Perceives the Zeppelins, 1924–1937," *South Atlantic Quarterly* 78:1 (Winter 1979).
2. I myself was standing the rudder watch for a sick crew member left behind in Pernambuco.

Chapter 7

1. Second round trip, June 9–19; fourth round trip, July 21–31; sixth round trip, August 18–28; eighth round trip, September 15–25; tenth round trip, October 13–23; twelfth round trip, December 8–19. All of these went via Pernambuco to Rio.

On the sixth and the twelfth trip, there were landings on the homeward leg at Seville, where there was a mooring mast.

Chapter 8

1. In German practice, this meant barometric pressure of 760 mm, relative humidity of 60 percent, air and gas temperatures of 0 degrees C, and hydrogen of specific gravity 0.1.

2. This contrasts with the figure usually seen of 105,000 cubic meters (3,707,550 cu ft).

3. Tambach Archives, Bockholt report to Reichs-Marine-Amt, Nov. 29, 1917.

4. Bockholt report.

5. Further, during the 1924 delivery flight of the *Los Angeles* to America, some 850,000 cubic feet of hydrogen, a good one-third of her content, had to be valved to compensate for the 29 tons of fuel and stores consumed during the three-day transatlantic crossing. For a regular passenger service this would be wasteful economically, and impractical logistically.

Chapter 10

1. The initial design was for a length of 811.1 feet with a finely tapered stern ending in a sharp point, even though the Germans knew that the interior length of the Lakehurst shed was only 804 feet. At a conference in Lakehurst in November 1934, Dr. Eckener was persuaded by Commander Rosendahl that the LZ 129 could not be housed diagonally in the shed, particularly with the *Los Angeles* sharing the space. It was then that he decided to bob and round off the tail cone.

2. Fuel capacity was 175,600 pounds, and water capacity 88,200 pounds.

3. The searchlight, used to take bearings at night, was of 5,200,000 candle power. It was mounted under the generator room in the keel, but was controlled from the control car.

4. Robert Jackson, *Airships* (New York: Doubleday, 1973), p. 189.

5. It must have been apparent to many people, however, that the sleeping cabins offered opportunities for a unique kind of daytime recreational activity. Webb Miller, the United Press foreign correspondent who was a passenger on the *Hindenburg's* first flight to the United States, wrote, "we discussed a suitable name for the first child conceived in mid-air aboard a Zeppelin—a possibility nowadays. I suggested Helium if it were a boy, and Shelium, if a girl." Webb Miller, *I Found No Peace* (New York: Simon and Schuster, 1936), p. 314; copyright 1936, 1963, by Webb Miller, reprinted by permission of Simon and Schuster, Inc.

6. The piano was not carried after the first few flights.

7. The *actual* weight empty of the *Hindenburg* on completion was 260,494 pounds. The *designed* weight empty was never revealed, but it is probable that she was a good 10 percent overweight.

Chapter 11

1. See Chapter 10, note 1.

2. Fickes actually arrived on March 20.

3. Pan American's three Martin 130 China Clipper flying boats, built to fly the Pacific from San Francisco to Manila via Hawaii, Midway, Wake, and Guam, were

the first airplanes to challenge Zeppelin dominance on the long ocean routes. While the normal range for the Martin Clippers was 3200 miles with twelve passengers, and 4000 miles with mail only and a crew of five, the longest leg (from San Francisco to Hawaii) was only 2400 miles. The first westward flight took off on November 22, 1935, and at the time I wrote, the *China Clipper* and *Philippine Clipper* were attempting to maintain a regular transpacific service, carrying mail only, but with rather frequent delays and several abortive starts. The first transpacific flight with paying passengers was not made until October 21, 1936, and then, frequently, the passenger load had to be reduced in favor of fuel. A seating capacity of fifty-two was just nominal, and possible only with minimum fuel. With a normal transpacific load of fourteen passengers, the Martin Clippers were uneconomical, and hardly appeared as a threat to the *Hindenburg*. Nor did their overnight sleeping accommodations match those of the airship.

4. *Deutsche Versuchsanstalt für Luftfahrt,* the German Aviation Experimental Center, which was the government licensing authority for civil aircraft.

5. E. A. Lehmann, *Auf Luftpatrouille und Weltfahrt* (Leipzig: Schmidt u. Günther, 1937), p. 360.

Chapter 12

1. A war hero and commander of naval fighting seaplanes based at Zeebrügge, Friedrich Christiansen later became an ardent Nazi. In 1935, he was made a director of the Zeppelin Reederei. In World War II, he was governor of Holland during the German occupation, and was later convicted of war crimes against the Dutch people.

2. The crew included the following: Commanders, 2. Watch officers, 3. Navigators, 4. Elevator men, 3. Helmsmen, 4. Radio operators, 4. Sailmakers, 3. Chief Engineer, 1. Assistant engineers, 2. Mechanics, 12. Keel watch, 3. Electricians, 3. Stewards, 7. Kitchen force, 3. *Total,* 54.

3. A spare Daimler was to have been shipped to Rio by sea before the first flight. Since this had not been done, it was carried on this flight and left there.

4. Engines 1 and 2 were the starboard and port after engines. Engines 3 and 4 were the starboard and port forward engines.

5. Used to help the elevator man to maintain a constant altitude, the statoscope, set to the altitude ordered by the watch officer, immediately showed the elevator man any deviation from the set altitude, which he was required to hold within ± 65 feet.

Chapter 13

1. This shed was completed in October 1938.

2. Everybody loved Willy, whose huge bulk and exuberant manner suggested a friendly bear. The son of a German nobleman and an American heiress, he loved to relate that at his christening in Berlin in 1904, Kaiser Wilhelm II had acted as his godfather.

Chapter 14

1. As Director General of Equipment (Luftzeugmeister), Udet test flew new and experimental combat aircraft with his old brilliance and dazzling skill, but as an administrator he was a disaster. As the war progressed and he was increasingly aware

of his failures, and was daily exposed to abuse and vituperation from Reichsmarschall Goering (who was jealous of Udet's World War I combat record), the time came when the sensitive and high-strung Udet could take no more, and on November 17, 1941, he committed suicide.

2. Admiral Miller's hook-on aircraft would have been one of two U.S. Navy Waco XJWs, civilian UBF three-place biplanes, bought as utility planes for the *Macon* in 1934. They were still in existence in August 1936, when the Bureau of Aeronautics ordered one of them re-equipped with a sky hook, but later the order was canceled.

Chapter 15

1. Sauter, whose landing station was in the bottom fin, was in no position to know if St. Elmo's fire was present on top of the ship. None was described at the inquiry by the official witnesses, who were standing more or less under the ship. Years later, however, Doug Robinson interviewed a couple who observed the disaster from some distance, standing outside the main gate of the air station, and who certainly saw St. Elmo's fire on top of the ship. They both noted a dim "blue flame" flickering along the backbone girder, and had time to exchange comments before there was a burst of flaming hydrogen from a point ahead of the upper fin.

2. On February 27, 1933, only a month after the Nazi seizure of power, the Reichstag building, home of the German parliament, was burned down. Undoubtedly Goering was the chief author of this act, which enabled Hitler to sweep away parliamentary government overnight. A half-witted Dutch Communist, Marius van der Lubbe, was seized and charged with the crime and after a public "show trial" was beheaded.

Chapter 17

1. Harold L. Ickes, *The Secret Diary of Harold L. Ickes.* Vol. 2, *The Inside Struggle 1936–39,* pp. 325 *et. seq.*

2. Hugo Eckener, *My Zeppelins,* translated by Douglas H. Robinson (London: Putnam, 1958), p. 180.

3. Many years later I learned that this was because the LZ 130 from her first trial flight carried secret electronic gear: Albert Sammt, *Mein Leben für den Zeppelin* (Stokach-Wahlwies: Verlag Pestalozzi Kinderdorf Wahlwies, 1982), p. 158.

Chapter 18

1. In 1917, two German Navy Zeppelins, L 49 and L 51, experienced serious though not fatal structural damage from implosion of the outer cover and underlying girder structure in descents from high altitudes. For the first time the outer cover was completely doped and airtight; with inadequate ventilation into the ship, increasing air pressure on the outer cover during descent caused major structural damage.

2. To get to the top of frame 17.5, and to install the reinforcements at the top fin attachments, the after gas cells would have had to be deflated, implying a layup of several weeks. This was to be postponed until after the completion of Fleet exercises.

Appendix C

1. "Heil Hitler!" with outstretched right arm.

2. In the *Graf Zeppelin* the crew space off the keel was called the "red tent," apparently because of the color of the light fabric serving as walls and ceiling of the compartment.

3. The first officer was the senior watch stander. The second officers commanded the other two watches. All three of these men held captain's ratings. The three third officers were navigators, serving in rotation with one of the three watch officers. Similarly, there were three persons to handle the duties of each crew position, standing watch in turn.

Index

Note: Italics indicate references in captions.